PRAISE FOR A WOR

"Thought-provoking in the extreme." —*Wired*

"Susskind's book is so timely, to miss it might be downright irresponsible." —*Booklist*

"Convincing and illuminating . . . A complex yet lucid and surprisingly optimistic account from the frontlines of technology addressing the challenges facing the human workforce." —*Kirkus Reviews*

"Fascinating and tightly argued." —*The Telegraph*

"One of the best things about this book is the way that Susskind will often offer excellent drive-by explanations of things he doesn't particularly need to explain, simply because they are interesting . . . If your plan is to get up to speed I would recommend it wholeheartedly." —*The Sunday Times* (London)

"A superb and sophisticated contribution to the debate over work in the age of artificial intelligence. Susskind approaches the discussion with a great command of the evidence and with excellent judgment. Never glib, consistently wise and well-informed, this is the book to read to understand how digital technologies and artificial intelligence in particular are reshaping the economy and labor market, and how we will live alongside increasingly smart machines."
—Jeffrey D. Sachs, author of *The End of Poverty*

"Susskind has written a fascinating book about a vitally important topic—and he writes with such elegance that you don't even notice how much you're learning. Original and compelling."
—Tim Harford, author of *The Undercover Economist*

"This is *the* book to read on the future of work in the age of artificial intelligence. It is thoughtful and state-of-the-art on the economics of the issue, but its real strength is the way it goes beyond just the economics. A truly important contribution that deserves widespread consideration."

—Lawrence H. Summers, former treasury secretary and director of the National Economic Council

"A pathbreaking, thought-provoking, and in-depth study of how new technology will transform the world of work."

—Gordon Brown, former prime minister of the United Kingdom

"Eloquent and humane, *A World Without Work* moves the debate beyond the illusion that technology always creates more jobs than it destroys. It provocatively explores the role of work in human life, and what to do when that role evaporates."

—Stuart Russell, author of *Human Compatible*

"Daniel Susskind offers an authoritative and hype-free perspective on how technology will change work. This eloquent and humane book deserves wide readership—and wide influence."

—Martin Rees, author of *On the Future*

"An important book on an equally important topic. Susskind's conclusion is that ultimately there will be less paid work to go around. This will shake the foundations of our economy and our society. It will be a daunting challenge. We have to start thinking hard about it now."

—Martin Wolf, author of *The Shifts and the Shocks*

Suki Dhanda

Daniel Susskind

A WORLD WITHOUT WORK

Daniel Susskind is a fellow in economics at Oxford University, a visiting professor at King's College London, and the coauthor of *The Future of the Professions*. Previously, he held various roles in the British government—in the Prime Minister's Strategy Unit, in the Policy Unit in 10 Downing Street, and in the Cabinet Office.

ALSO BY DANIEL SUSSKIND

The Future of the Professions:
How Technology Will Transform the Work of Human Experts
(with Richard Susskind)

A WORLD
WITHOUT WORK

A WORLD WITHOUT WORK

Technology, Automation, and How We Should Respond

DANIEL SUSSKIND

PICADOR

METROPOLITAN BOOKS
HENRY HOLT AND COMPANY NEW YORK

Picador
120 Broadway, New York 10271

Printed in the United States of America
Originally published in 2020 by Metropolitan Books / Henry Holt and Company
First Picador paperback edition, 2021

The Library of Congress has cataloged the Metropolitan Books hardcover edition as follows:
Names: Susskind, Daniel, author.
Title: A world without work : technology, automation, and how we should respond / Daniel Susskind.
Description: First edition. | New York, N.Y. : Metropolitan Books/Henry Holt & Company, 2020. | Includes bibliographical references and index. |
Identifiers: LCCN 2019029199 | ISBN 9781250173515 (hardcover) | ISBN 9781250173522 (ebk)
Subjects: LCSH: Automation—Social aspects. | Technology—Social aspects. | Social change.
Classification: LCC HD6331.S86 2020 | DDC 331.25—dc23
LC record available at https://lccn.loc.gov/2019029199

Picador Paperback ISBN: 978-1-250-80825-7

Designed by Kelly S. Too

Our books may be purchased in bulk for promotional, educational, or business use. Please contact your local bookseller or the Macmillan Corporate and Premium Sales Department at 1-800-221-7945, extension 5442, or by email at MacmillanSpecialMarkets@macmillan.com.

For book club information, please visit facebook.com/picadorbookclub or email marketing@picadorusa.com.

picadorusa.com • instagram.com/picador
twitter.com/picadorusa • facebook.com/picadorusa

1 3 5 7 9 10 8 6 4 2

For Grace and Rosa

CONTENTS

PREFACE TO THE PAPERBACK EDITION

This book is about one of the greatest economic challenges of our time: the threat of a world where there is not enough well-paid work for everyone to do, because of the remarkable technological changes on the horizon. It was written with a sense of urgency, since in my view we are not yet taking this threat seriously enough. But nobody could have predicted how, only a few months after this book was first published, a global pandemic would bring economic life as we knew it to an end and make the ideas and concerns in here more urgent than ever.

At the time of this writing, the COVID-19 pandemic has been with us for about six months. When the pandemic began, the hope was that it would be a short-lived crisis. Economies would need to be temporarily placed in a sort of suspended animation, but once the virus had passed—in a matter of weeks, it was originally thought—we would swiftly return to economic life as usual. As we now know, though, this initial hope turned out to be completely misplaced. The virus is here to stay for some time. The frantic policy firefighting of the first few months of the crisis has given way to far more long-lasting interventions. And the economic consequences of the pandemic have been more destructive than most of us had first imagined. From April to June 2020, for instance, the US suffered the sharpest collapse in output since World War II. The UK lost almost eighteen years of growth in a matter of months.[1]

At the core of this economic meltdown has been the labor market. Work was already precarious in many parts of the world before the pandemic began, marked by stagnating wages, rising insecurity, pockets of unemployment, and declining participation. COVID-19 pushed it off a cliff: in many countries hit hard by the virus, like the US and the UK, joblessness has surged to extraordinary levels. As the pandemic took hold, in other words, we found ourselves unexpectedly thrust into a world with much less work—not because that work has been automated, but because the measures that we were forced to adopt in response to the virus (lockdowns, social distancing, self-isolation, and so on) completely decimated the demand that so many jobs rely upon.

As a result, we have had to confront even sooner the challenges that concern me in this book. Andrew Yang, the 2020 US presidential candidate who focused on the threat of job displacement, made this point well: "Apparently I should have been talking about a pandemic instead of automation," he wrote on Twitter.[2] To be clear, the threat of technological unemployment in the future has not diminished; on the contrary, there are reasons to think that the threat is now greater than before. But the pandemic has also given us a frightening preview of what this future might look like, and an insight into the immensity of the challenges that we will have to face when it arrives.

A GLIMPSE OF THE FUTURE

As we shall see in this book, the fundamental difficulty that lies ahead is a distributional one. Technological progress may make us collectively more prosperous than ever before, but how are we to share out that prosperity when our traditional way of doing so—paying a wage for the work that people do—is less effective than in the past? And this, of course, is precisely the economic problem that has dominated in 2020. Overnight, vast numbers of workers around the world woke up to suddenly find themselves without a job or an income.

What should be done? I argue that in a moment like this the state must take on a far larger role in sharing out prosperity in society, through what I call the Big State. The pandemic has now proven that there is no credible alternative. Different countries have adopted slightly different schemes, but all of them involve a vastly bigger state providing an income to those without work. Indeed, ideas that only a few months ago were viewed by some as outlandish—a basic income, for instance—have swiftly become commonplace in almost all corners of political conversation. To provide support to the unemployed, and to prop up the economy more generally, the US has already borrowed more than five times what it did at the height of the financial crisis of 2007–8; the UK is on track in 2020 to set a peacetime borrowing record.[3]

Aside from sharing out prosperity, there are two other big challenges we can expect to face in a world with less work, both of which have little to do with economics. One of these is the growing power of a handful of large technology companies, or Big Tech. Here, too, the pandemic offers a glimpse of the future: it is a conspicuous feature of the COVID-19 economic landscape that such companies have done particularly well. At one point during this crisis, just five of them accounted for more than 20 percent of the worth of the entire S&P 500 index, which comprises five hundred large companies listed on US stock exchanges.[4] Apple alone was worth more than all the companies in the London Stock Exchange's FTSE 100 index combined.[5]

My concern in this book, though, is far less with the tech companies' economic power—great and growing though it may be—than with their *political* power, and the impact they may have on issues of liberty, democracy, and social justice in the future. So it is important to note, for instance, how debates about data privacy and security have quietly disappeared from public discussion since the pandemic began. A "do whatever it takes" mentality took hold at the start of this crisis, with many countries permitting CCTV surveillance footage, smartphone location data, and credit card purchasing history, among much else, to be collected, sifted, sorted, and studied on a huge scale in an effort to control the virus. The threat may have required this. But in time we must ensure that the new political power we have granted to Big Tech, and the heightened ability to shape how we all live together in society that comes with it, is properly scrutinized and reined back in if need be.

The final challenge that we will face in a world with less work, I argue, is finding meaning in life. It is often said that work is not simply a source of income but a source of purpose as well. And so, if employment dries up, where will that sense of direction come from? My own view is that the relationship between work and meaning is actually far murkier than is commonly supposed: many people do not get a strong sense of purpose from their jobs today, and our relationship to work has looked wildly different at other moments in history. The pandemic has strengthened that belief. Yes, there have been awful stories of those who lost work and felt a sense of devastation that could not be explained by the loss of an income alone; but there are also many accounts of those who felt quite the opposite, a sense of relief in being freed from jobs that were simply not worth the wages they provided.

But what will people actually *do* if they do not have to work for a living? I fear we do not yet have good answers to this question. In a world like ours, where work sits at the center of our lives, it is very difficult to imagine how we might spend our time any differently. Our struggles in this pandemic-induced world have shown that. We can point to some telling changes in spending patterns over the last few months: the UK, for instance, has suffered major shortages of flour, timber, and bedding plants as people took up baking, home-improvements, and gardening to fill their spare time; the US has suffered from similarly disruptive spikes in demand. But there have also been unfamiliar public conversations about bigger issues: work-life balance, the value of family and community, the merits of city life, how best to spend our time in idleness, how to maintain our mental health in tough times. (Depression among UK adults almost doubled at the start of the pandemic, while text messages to a US government mental health hotline rose almost tenfold.)[6] That these conversations feel so novel, and that the conclusions can at times seem so provisional and unsatisfying, strengthens my sense that the all-consuming nature of our traditional working lives has distracted us from these big questions until now.

THE INCREASING THREAT OF AUTOMATION

While the pandemic offers a preview of the problems that a world with greater automation will have to grapple with—issues regarding the distribution of prosperity, the power of Big Tech, and the search for meaning—it is also likely hastening the arrival of that world.

One reason for this is that many countries around the globe are now in severe recessions, and evidence from the past suggests that when economies slow down, automation can pick up. Around the turn of the twenty-first century, for example, jobs for secretaries, clerks, salespeople, and the like shriveled up (as a proportion of total employment) as new technologies began to take on these workers' tasks and displace them from their roles. In the book, I explore exactly why such "middle-skilled" jobs were lost while both high-paid and low-paid workers increased their share of employment. But what matters for thinking about our current situation is that, at least in the US, the vast majority of this job destruction took place during economic downturns. One influential study suggests that since the mid-1980s, 88 percent of these middle-skill job losses took place within a year of a recession.[7]

What's more, our particular downturn is no ordinary recession. The pandemic also creates new and unique reasons to worry about the threat of automation. Most obviously, COVID-19 strengthens the incentive to replace human beings with machines. A machine, after all, will not pass the virus on to coworkers or customers; it will not fall ill and need to take time off from work; it will not need to isolate to protect its peers.

So far, this incentive has been kept in check to a certain extent by government interventions. The UK government, for instance, was at one point paying up to 80 percent of the wages of 9.6 million workers—more than a third of all UK employees—to protect them from unemployment.[8] But many governments have not taken this route. And when those interventions that do exist are relaxed—as they inevitably will be—the incentive to automate will grow even stronger. For businesses looking to boost productivity during the downturn or cut labor costs as revenue falls, replacing workers with machines for particular activities might seem increasingly attractive. At the start of the pandemic, for instance, a survey of global business executives by the consulting firm EY found that 41 percent were investing in accelerating automation.[9]

Finally, the pandemic may have softened some of the cultural resistance that accompanies the use of new technologies in the workplace. The barriers to automation, after all, are not simply technological ("Is it possible to automate a task?"), economic ("Is it profitable to automate a task?"), or regulatory ("Is it permitted to automate a task?"). They are also cultural: what is automated depends, in part, on whether people find it palatable to do something with a machine. And to the extent that any of us—whether as business owners, employers, employees, or consumers—might have had a bias against new technologies before the pandemic, this crisis is likely to have weakened it. One poll, for instance, suggests that all age groups in Britain now "feel more positively" toward technology; another, that a third of Brits have become "more confident in using technology" as well.[10] Out of necessity, we have been forced to use technology in ways that would have simply seemed unimaginable a few months ago—and it has largely been a success. And so, any particular act of automation in the future is now likely to feel far less like an unprecedented leap.

Take medicine, for example. Before the pandemic began, about 80 percent of doctors' appointments in England and Wales were conducted face-to-face; now, that proportion has fallen to only about 7 percent.[11] It is hard to believe that virtual appointments will stop once the pandemic is over; on the other hand, it is easy to imagine that other parts of medicine—diagnostics, for instance—could also start being done differently through technology, and perhaps without involving doctors at all. Or take the law. In many jurisdictions, brick-and-mortar courtrooms were shuttered, and practically overnight court became an online service rather than a physical place. As with medicine, not only is it possible to see how such a virtual setup might now become the norm in certain corners of the criminal justice system, but bolder technological proposals—for instance, that some cases, such as low-value civil disputes, might be settled without any human deliberation at all—seem far less radical than they would have only a few months ago.

LOWER-PAID WORKERS AT RISK

At the moment, to be sure, technology mostly appears to be keeping people in work rather than pushing them out. Many have been able to use technology to work remotely, to an extent that would have seemed unimaginable until recently: in the US and UK, as the crisis began, about two-thirds of those still working were remote.[12] However, not everyone can work from home, and those who can tend to be in better-paid, white-collar roles. A US survey found that while 71 percent of people earning more than $180,000 a year could work remotely during the pandemic, only 41 percent of those earning less than $24,000 could. Another study reported that while 62 percent of workers with a bachelor's degree or more could perform their jobs from home, only 9 percent of those who did not graduate from high school could do so.[13] Remote work is simply not an option for many blue-collar workers, such as those who work in restaurants, shops, and warehouses.

This particular inequality in the ability of workers to adapt to the pandemic through technology is symptomatic of a deeper problem. When the crisis began, it was said that COVID-19 would be a "great leveler." This disease, many proclaimed, would not discriminate according to a person's ethnicity or wealth; all of us were equally at risk. We now know that this was a myth. To start with, the medical impact of the virus has been extremely unequal. In the UK, people from ethnic minority backgrounds made up 14 percent of the population but 34 percent of critically ill COVID-19 patients; in the US, Black people were almost five times more likely to be hospitalized due to the virus, and more than twice as likely to die from it, than white people.[14] And the *economic* impact of the virus has been extremely unequal, too. The job losses, for instance, have been concentrated among the lower-paid workers: one study suggests that, in the US, workers in the bottom 20 percent of earners were about four times more likely to lose their job at the start of the pandemic than those in the top 20 percent of earners.[15]

These inequalities are striking in themselves, but they are also important for thinking about the looming threat of automation. The pandemic has both likely increased that threat and made it clear that workers who are already economically disadvantaged may be hardest hit.

Over the past few decades, lower-paid workers have mostly been protected from automation. This is because their jobs very often involve personal interaction or manual labor, and until recently these tasks have proven tricky to automate. But the cruel irony of the last few months is that these workers have in fact been hardest hit by the pandemic precisely *because* of those properties of their jobs: the virus spreads through personal interaction, and it flourishes in poorly ventilated indoor spaces such as factories and warehouses. As a result, many of these people have found themselves unable to work.

As the pandemic has increased the incentive to automate, therefore, it is likely that these hands-on workers are the most at risk: they cannot readily work in their traditional workplaces, nor can they retreat to a home office and do their job from there. It is no surprise that so many recently reported technological developments seem aimed directly at replacing them: machines that stock shelves, prepare packages, greet customers, deliver goods, clean floors, take temperatures, and so on.

Does the prospect of an effective vaccine mean that this incentive to automate, however strong it may be at the moment, will eventually fizzle out? That may indeed be the case. But it is not clear that such a development—however magnificent it would be from a medical point of view—would bring the threat of automation back down. To begin with, the cultural shifts mentioned before may persist: if the pandemic has made us more welcoming toward technology, that new attitude will probably remain. More significantly, the pandemic has also transformed the fundamental rhythms of how many of us live our lives: we eat out less, shop online more, avoid travel if we can, stay away from theaters and cinemas and sporting events, work from home if possible, and so on. Even when the pandemic fades away and government restrictions are relaxed, these changes in habits and behaviors are unlikely to reverse completely.[16]

Those who say that the pandemic spells the "end of the office," the "death of Main Street," or the "collapse of the city center" are probably overstating their case. Though offices and shopping destinations were abandoned for a while, people are slowly starting to return.[17] Nevertheless, it is entirely plausible that such places will remain diminished versions of their former selves for quite some time—perhaps indefinitely. And if that is right, it does not bode well for workers who depend on these places: security guards, receptionists, and cleaners in offices; waiters, sandwich makers, and baristas in nearby streets; retailers, transport workers, hotel staff, and entertainers in city centers; and so on. In this scenario, of course, the decline in demand for their work may be due more to the effects of the pandemic than to technology as such. But when thinking about the threat of automation, it is crucial to consider these shifts, because the lower-paid, hands-on jobs are precisely the ones that provided people displaced by machines with work in the past—and their future is now in doubt.

In a sense, the pandemic has been a pilot scheme in how we ought to respond to a world with less work. This exercise has been unplanned and unwanted, but it has also proven to be informative and revelatory. I hope that in the months and years to come we are able to reflect on this vast social experiment, to understand what has worked in responding to this crisis, and to be honest about where we have fallen short. At the moment, we are only temporary visitors in a world with less work. This pandemic, like all those before it, will eventually pass, and the problems that consume us today will fall away. But when the COVID-19 crisis recedes, the threat of automation may have only increased. And then the challenges that we have caught an unsettling glimpse of during the pandemic will reemerge, and start to trouble and test us once again.

Daniel Susskind
London, England
30 September 2020

A WORLD WITHOUT WORK

Introduction

The "Great Manure Crisis" of the 1890s should have come as no surprise.[1] For some time, in big cities like London and New York, the most popular forms of transport had relied upon horses—hundreds of thousands of them—to heave cabs, carts, wagons, wains, and a variety of other vehicles through the streets. As locomotives, horses were not particularly efficient: they had to take a break to rest and recover every few miles, which partly explains why quite so many were needed.[2] Operating a basic carriage, for example, required at least three animals: two working in rotation to pull it along, plus one in reserve in case of a breakdown. The horse-drawn tram, the transit mode of choice for New Yorkers, relied on a team of eight, which took turns dragging it on a set of specially laid tracks. And in London, thousands of horse-drawn double-decker buses, modestly sized versions of today's red ones, demanded about a dozen animals apiece for the task.[3]

With these horses came manure—and lots of it. A healthy horse produces somewhere between fifteen and thirty pounds of manure a day, almost the weight of a two-year-old child.[4] One enthusiastic health officer working in Rochester, New York, calculated that the horses in his city alone produced enough in a year to cover an acre of land to a height of 175 feet, almost as high as the Leaning Tower of Pisa.[5] Apocryphally, people at the time extrapolated from these calculations to an inescapably manure-filled future: a New York commentator who predicted that piles would soon reach the height of third-story windows, a London reporter who imagined that by the middle of the twentieth century the streets would be buried under nine feet of the stuff.[6] Nor was the crisis simply about manure. Thousands of putrefying dead horses littered the roads, many deliberately left to decay to a size that made for easier disposal. In 1880 alone, about fifteen thousand horse carcasses were removed from New York City.[7]

It is said that policymakers did not know what to do.[8] They couldn't simply ban horses from the streets: the animals were far too important. In 1872, when the so-called Horse Plague hit the United States, with horses struck down by one of the worst outbreaks of equine flu in recorded history, large parts of the country's economy came to a halt.[9] Some even blame the epidemic for that year's Great Fire of Boston; seven hundred buildings burned to the ground, they claim, because there were not enough horses to pull firefighting equipment to the scene.[10] But the twist in the tale is that, in the end, policymakers didn't need to worry. In the 1870s, the first internal combustion engine was built. In the 1880s, it was installed in the first automobile. And only a few decades later, Henry Ford brought cars to the mass market with his famous Model T. By 1912, New York had more cars than horses. Five years after that, the last horse-drawn tram was decommissioned in the city.[11] The Great Manure Crisis was over.

The "Parable of Horseshit," as Elizabeth Kolbert called it in the *New Yorker*, has been told many times over the years.[12] In most versions of the story, the decline of horses is cast in an optimistic light, as a tale of technological triumph, a reassuring reminder that it is important to remain open-minded even when you find yourself knee-deep in a foul, seemingly intractable problem. But for Wassily Leontief, the Russian-American economist who won the Nobel Prize in 1973, the same events suggested a more unsettling conclusion. What he saw instead was how a new technology, the combustion engine, had taken a creature that, for millennia, had played a central role in economic life—not only in cities but on farms and fields—and, in only a matter of decades, had banished it to the sidelines. In a set of articles written in the early 1980s, Leontief made one of the most infamous claims in modern economic thought. What technological progress had done to horses, he said, it would eventually do to human beings as well: drive us out of work. What cars and tractors were to them, he thought, computers and robots would be to us.[13]

Today, the world is gripped again by Leontief's fear. In the United States, 30 percent of workers now believe their jobs are likely to be replaced by robots and computers in their lifetime. In the UK, the same proportion think it could happen in the next twenty years.[14] And in this book, I want to explain why we have to take these sorts of fears seriously—not always their substance, as we shall see, but certainly their spirit. Will there be enough work for everyone to do in the twenty-first century? This is one of the great questions of our time. In the pages that follow, I will argue that the answer is "no" and explain why the threat of "technological unemployment" is now real. I will describe the different problems this will create for us—both now and in the future—and, most important, set out how we might respond.

It was John Maynard Keynes, the great British economist, who popularized the term "technological unemployment" almost fifty years before Leontief wrote down his worries, capturing in a pithy pairing of words the idea that new technologies might push people out of work. In what follows, I will draw on many of the economic arguments that have been developed since Keynes to try to gain a better look back at what happened in the past, and a clearer glimpse of what lies ahead. But I will also seek to go well beyond the narrow intellectual terrain inhabited by most economists working in this field. The future of work raises exciting and troubling questions that often have little to do with economics: questions about the nature of intelligence, about inequality and why it matters, about the political power of large technology companies, about what it means to live a meaningful life, about how we might live together in a world that looks very different from the one in which we have grown up. In my view, any story about the future of work that fails to engage with these questions as well is incomplete.

NOT A BIG BANG, BUT A GRADUAL WITHERING

An important starting point for thinking about the future of work is the fact that, in the past, many others have worried in similar ways about what lies ahead—and been very wrong. Today is not the first time that automation anxiety has spread, nor did it first appear in the 1930s with Keynes. In fact, ever since modern economic growth began, centuries ago, people have periodically suffered from bouts of intense panic about being replaced by machines. Yet those fears, time and again, have turned out to be misplaced. Despite a relentless flow of technological advances over the years, there has always been enough demand for the work of human beings to avoid the emergence of large pools of permanently displaced people.

And so, in the first part of the book, I begin with this history, investigating why those who worried about being replaced by machines turned out repeatedly to be so wrong, and exploring how economists have changed their minds over time about the impact of technology on work. Then I turn to the history of artificial intelligence (AI)—a technology that has captured our collective imagination over the last few years, and which is largely responsible for the renewed sense of unease that many now feel about the future. AI research, in fact, began many decades ago, with an initial burst of enthusiasm and excitement, but that was followed by a slump into a long, deep winter when little progress was made. In recent years, though, there has been a rebirth, an intellectual and practical revolution that caught flat-footed many economists, computer scientists, and others who had tried to predict which activities machines could never do.

In the second part of the book, building on this history, and trying to sidestep the intellectual mistakes that others have made before, I explain how technological unemployment is likely to unfold in the twenty-first century. In a recent survey, leading computer scientists made the claim that there is a 50 percent chance that machines will outperform human beings at "every task" within forty-five years.[15] But the argument I make does not rely on dramatic predictions like this turning out to be true. In fact, I find it hard to believe that they will. Even at the century's end, tasks are likely to remain that are either hard to automate, unprofitable to automate, or possible and profitable to automate but which we will still prefer people to do. And despite the fears reflected in those polls of American and British workers, I also find it difficult to imagine that many of today's jobs will vanish completely in years to come (to say nothing about new types of jobs that await in the future). Much of that work, I expect, will turn out to involve some tasks that lie beyond the reach of even the most capable machines.

The story I tell is a different one. Machines will not do everything in the future, but they will do *more*. And as they slowly, but relentlessly, take on more and more tasks, human beings will be forced to retreat to an ever-shrinking set of activities. It is unlikely that every person will be able to do what remains to be done; and there is no reason to imagine there will be enough demand for it to employ all those who are indeed able to do it.

In other words, if you picked up this book expecting an account of a dramatic technological big bang in the next few decades, after which lots of people suddenly wake up to find themselves without work, you will be disappointed. That scenario is not likely to happen: some work will almost certainly remain for quite some time to come. But, as time passes, that work is likely to sit beyond the reach of more and more people. And, as we move through the twenty-first century, the demand for the work of human beings is likely to wither away, gradually. Eventually, what is left will not be enough to provide everyone who wants it with traditional well-paid employment.

A useful way of thinking about what this means is to consider the impact that automation has had already had on farming and manufacturing in many parts of the world. Farmers and factory workers are still needed: those jobs have not completely vanished. But the number of workers that is needed has fallen in both cases, sometimes precipitously—even though these sectors produce more output than ever before. There is, in short, no longer enough demand for the work of human beings in these corners of the economy to keep the same number of people in work. Of course, as we shall see, this comparison has its limits. But it is still helpful in highlighting what should actually be worrying us about the future: not a world without *any* work at all, as some predict, but a world without *enough* work for everyone to do.

There is a tendency to treat technological unemployment as a radical discontinuity from economic life today, to dismiss it as a fantastical idea plucked out of the ether by overly neurotic shock-haired economists. By exploring how technological unemployment might actually happen, we will see why that attitude is a mistake. It is not a coincidence that, today, worries about economic inequality are intensifying at the exact same time that anxiety about automation is growing. These two problems—inequality and technological unemployment—are very closely related. Today, the labor market is the main way that we share out economic prosperity in society: most people's jobs are their main, if not their only, source of income. The vast inequalities we already see in the labor market, with some workers receiving far less for their efforts than others, show that this approach is already creaking. Technological unemployment is simply a more extreme version of that story, but one that ends with some workers receiving nothing at all.

In the final part of the book, I untangle the different problems created by a world with less work and describe what we should do about them. The first is the economic problem just mentioned: how to share prosperity in society when the traditional mechanism for doing so, paying people for the work that they do, is less effective than in the past. Then I turn to two issues that have little to do with economics at all. One is the rise of Big Tech, since, in the future, our lives are likely to become dominated by a small number of large technology companies. In the twentieth century, our main worry may have been the economic power of corporations: but in the twenty-first, that will be replaced by fears about their *political* power instead. The other issue is the challenge of finding meaning in life. It is often said that work is not simply a means to a wage but a source of direction: if that is right, then a world with less work may be a world with less purpose as well. These are the problems we will face, and each of them will demand a response.

A PERSONAL STORY

The stories and arguments in this book are, to some extent, personal ones. About a decade ago, I began to think about technology and work in a serious way. Well before this, however, it had been an informal interest, something I often mulled over. My father, Richard Susskind, had written his doctorate in the 1980s at Oxford University on artificial intelligence and law. During those years, he had squirreled himself away in a computing laboratory, trying to build machines that could solve legal problems. (In 1988, he went on to codevelop the world's first commercially available AI system in law.) In the decades that followed, his career built upon this work, so I grew up in a home where conundrums about technology were the sorts of things we chewed over in dinner-table conversation.

When I left home, I went to Oxford to study economics. And it was there, for the first time, that I was exposed to the way that economists tend to think about technology and work. It was enchanting. I was enthralled by the tightness of their prose, the precision of their models, the confidence of their claims. It seemed to me that they had found a way to strip away the disorienting messiness of real life and reveal the heart of the problems.

As time passed, my initial enchantment dulled. Eventually, it disappeared. After graduating, I joined the British government—first in the Prime Minister's Strategy Unit, then in the Policy Unit in 10 Downing Street. There, buoyed by technologically inclined colleagues, I started to think more carefully about the future of work and whether the government might have to help in some way. But when I turned for help to the economics I had learned as an undergraduate, it was far less insightful than I had hoped. Many economists, as a matter of principle, want to anchor the stories they tell in past evidence alone. As one eminent economist put it, "Although we all enjoy science fiction, history books are usually a safer guide to the future."[16] I was not convinced by this sort of view. What was unfolding in the economy before me looked radically different from experiences of what had come before. I found this very disconcerting.

And so, I left my role in British government and, after time spent studying in America, returned to academia to explore various questions about the future of work. I completed a doctorate in economics, challenging the way that economists had traditionally thought about technology and work, and tried to devise a new way to think about what was happening in the labor market. At the same time, I teamed up with my father to write *The Future of the Professions*, a book that explored the impact of technology on expert white-collar workers—lawyers, doctors, accountants, teachers, and others. When we began our research for that project a decade ago, there was a widespread presumption that automation would only affect blue-collar workers. It was thought that professionals were somehow immune from change. We challenged that idea, describing how new technologies would allow us to solve some of the most important problems in society—providing access to justice, keeping people in good health, educating our children—without relying on traditional professionals as we had done in the past.[17]

Insights from both my academic research and our book on the professions will reappear in the pages that follow, sanded into better shape through subsequent experience and thinking. In short, then, this book captures my own personal journey, a decade spent thinking almost entirely about one particular issue—the future of work.

GOOD PROBLEMS TO HAVE

Although these opening words may suggest otherwise, this book is optimistic about the future. The reason is simple: in decades to come, technological progress is likely to solve the economic problem that has dominated humanity until now. If we think of the economy as a pie, as economists like to do, the traditional challenge has been to make that pie large enough for everyone to live on. At the turn of the first century AD, if the global economic pie had been divided into equal slices for everyone in the world, each person would have received just a few hundred of today's dollars per year. Most people lived around the poverty line. Roll forward one thousand years, and roughly the same would have been true. Some even claim that, as late as 1800, the average person was no more materially prosperous than her equivalent back in 100,000 BC.[18]

But over the last few hundred years, economic growth has soared, and this growth was driven by technological progress. Economic pies around the world have become much bigger. Today, global GDP per capita, the value of those equally sized individual slices, is already about

$10,720 a year (an $80.7 trillion pie shared out among 7.53 billion people).[19] If economies continue to grow at 2 percent per year, our children will be twice as rich as us. If we expect a measlier 1 percent annual growth, then our grandchildren will be twice as well off as we are today. We have, at least in principle, come very close to solving the problem that plagued our fellow human beings in the past. As the economist John Kenneth Galbraith so lyrically put it, "man has escaped for the moment the poverty which was for so long his all-embracing fate."[20]

Technological unemployment, in a strange way, will be a symptom of that success. In the twenty-first century, technological progress will solve one problem, the question of how to make the pie large enough for everyone to live on. But, as we have seen, it will replace it with three others: the problems of inequality, power, and purpose. There will be disagreement about how we should meet these challenges, about how we should share out economic prosperity, constrain the political power of Big Tech, and provide meaning in a world with less work. These problems will require us to engage with some of the most difficult questions we can ask—about what the state should and should not do, about the nature of our obligations to our fellow human beings, about what it means to live a meaningful life. But these are, in the final analysis, far more attractive difficulties to grapple with than the one that haunted our ancestors for centuries—how to create enough for everyone to live on in the first place.

Leontief once said that "if horses could have joined the Democratic party and voted, what happened on farms might have been different."[21] It is a playful phrase with a serious point. Horses did not have any control over their collective fate, but we do. I am not a technological determinist: I do not think the future must be a certain way. I agree with the philosopher Karl Popper, the enemy of those who believe that the iron rails of our fate have already been set down for us to trundle along, when he says that "the future depends on ourselves, and we do not depend on any historical necessity."[22] But I am also a technological realist: I do think that our discretion is constrained. In the twenty-first century, we will build systems and machines that are far more capable than those we have today. I don't believe we can escape that fact. These new technologies will continue to take on tasks that we thought only human beings would ever do. I do not believe we can avoid that, either. Our challenge, as I see it, is to take those unavoidable features of the future as given, and still build a world where all of us can flourish. That is what this book is about.

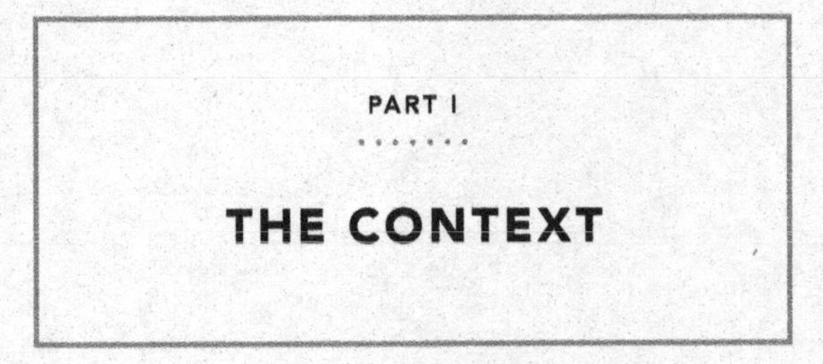

PART I

THE CONTEXT

1

A History of Misplaced Anxiety

Economic growth is a very recent phenomenon. In fact, for most of the three hundred thousand years that human beings have been around, economic life has been relatively stagnant. Our more distant ancestors simply hunted and gathered what little they needed to survive, and that was about it.[1] But over the last few hundred years, that economic inactivity came to an explosive end. The amount each person produced increased about thirteen-fold, and world output rocketed nearly three hundred-fold.[2] Imagine that the sum of human existence was an hour long: most of this action happened in the last half-second or so, in the literal blink of an eye.

Economists tend to agree with one another that this growth was propelled by sustained technological progress, though not on the reasons why it started just where and when it did—in Western Europe, toward the end of the eighteenth century.[3] One reason may be geographical: certain countries had bountiful resources, a hospitable climate, and easily traversable coastlines and rivers for trade. Another may be cultural: people in different communities, shaped by very different intellectual histories and religions, had different attitudes toward the scientific method, finance, hard work, and each other (the level of "trust" in a society is said to be important). The most common explanation of all,

Figure 1.1: Global Output Since AD 1[4]

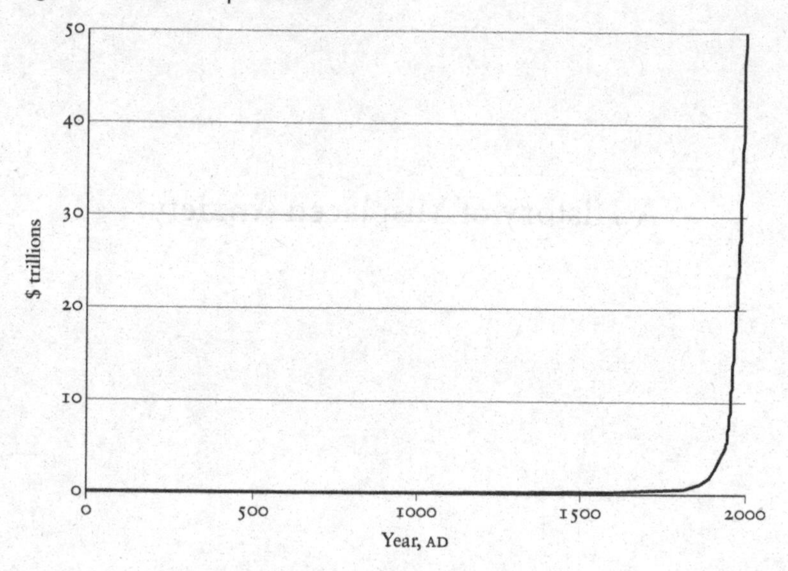

though, is institutional: certain states protected property rights and enforced the rule of law in a way that encouraged risk taking, hustle, and innovation, while others did not.

Whatever the particular reasons, it was Britain that led the economic charge, thundering ahead of others in the 1760s.[5] Over the following decades, new machines were invented and put to use that greatly improved the way that goods were produced. Some, like the steam engine, have become standard symbols of economic progress and technological ingenuity. And dramatic as the term "revolution" may seem, it is probably still an understatement: the Industrial Revolution is one of the most significant moments in the history of humankind. Before this period, any economic growth had been limited, stuttering, and quickly fizzled out. Afterward, it started to flow relatively bountifully and steadily. Today, we have become entirely dependent upon this economic fix. Think of the eruptions of anger and anxiety, the waves of frustration and despondency that crash through society each time economic growth stops or even slows. It is as if we can no longer live well without it.

The new technologies of the Industrial Revolution allowed manufacturers to operate more productively than ever before—in short, to make far more with far less.[6] And it is here, at the beginning of modern economic growth, that we can also detect the origins of "automation anxiety." People started to worry that using these machines to make more things would also mean less demand for their own work. From the outset, it seems, economic growth and automation anxiety were intertwined.

Of course, people must have been anxious about automation even before then. For any invention, it is possible to imagine or identify some group of unlucky people who might have felt threatened. The printing press, for instance—perhaps the most consequential of all technologies predating the Industrial Revolution—was initially met with resistance from human scribes who wanted to protect their traditional craft. Regarding printed Bibles, they said that only the devil himself could produce so many copies of a book so swiftly.[7] But the particular character of the changes that took place during the Industrial Revolution was different from the past. Their intensity, breadth, and persistence gave a fresh severity to the familiar worries.

AUTOMATION ANXIETY

This anxiety that automation would destroy jobs spilled into protest and dissent. Consider the experience of James Hargreaves, the modest man who invented the spinning jenny. An illiterate cotton weaver, he retreated to a remote village in Lancashire, England, to build his device in peace. This was a machine that would allow thread to be spun from cotton far more swiftly than with human hands alone, a valuable innovation at a time when turning raw cotton into useable thread was a growing business. (In fact, by the middle of the nineteenth century, Britain would be producing half of all the world's cloth.)[8] But when word spread about what Hargreaves was up to, his neighbors broke in, demolished the machine, and, somewhat gratuitously, destroyed his furniture, too. When Hargreaves tried to set up a factory elsewhere, he and his business partner were set upon by a mob.[9]

John Kay, a contemporary of Hargreaves, appears to have suffered a similar fate when he invented the flying shuttle in the 1730s. His home, it is said, was likewise ransacked by furious weavers, who "would have killed him had he not been conveyed to a place of safety by two friends in a wool-sheet."[10] A nineteenth-century mural in the town hall in Manchester, England, depicts his surreptitious flight from danger.[11]

These were not isolated incidents. During the Industrial Revolution, such technological vandalism was widespread. As is now very well-known, these marauders were called "Luddites." They took their name from Ned Ludd, an apocryphal East Midlands weaver who smashed a set of framing machines at the start of the Industrial Revolution. Ned was probably not a real person, but the disturbances his flag-bearers caused certainly were. In 1812, the British Parliament felt forced to pass the "Destruction of Stocking Frames, etc. Act." Destroying machines became a crime punishable by death, and several people were soon charged and executed. The following year, the punishment was softened to deportation to Australia—but that, it turned out, was not sufficiently unpleasant, and death was reinstated as the penalty in 1817.[12] Today, we still call our technologically disinclined contemporaries Luddites.

Before the Industrial Revolution, the state was not always on the side of the inventors. Indeed, there were moments when it was so troubled by the discontent of disgruntled workers that it stepped in and tried to stop the offending innovations from spreading. Consider two stories from the 1580s. First there was William Lee, an English priest, who invented a machine to free people from having to knit with their hands. In 1589, he made his way to London, hoping to show his invention to Queen Elizabeth I and get a patent to protect it. But when she saw the machine she flat-out refused, replying, "Thou aimest high, Master Lee. Consider thou what the invention could do to my poor subjects. It would assuredly bring to them ruin by depriving them of employment, thus making them beggars."[13] Then there is the tragedy of Anton Möller, who had the bad luck to invent the ribbon loom in 1586—bad luck because rather than simply refuse a patent, the city council in his hometown of Danzig is said to have responded to his triumph with an order that he be strangled, hardly the warm reception that we reserve for today's entrepreneurs.[14]

But it was not only workers and the state who were anxious. As time passed, economists also started to take the threat of automation seriously. As noted before, it was Keynes who would popularize the term "technological unemployment" in 1930. But David Ricardo, one of the founding fathers of economics, struggled with this issue more than a century before him. In 1817, Ricardo published his great work, *Principles of Political Economy and Taxation*. Within four years of publication, though, he released a fresh edition with a new chapter, "On Machinery." In it, he made a significant intellectual concession, declaring that he had changed his mind on the question of whether technological progress would benefit workers. His lifelong assumption that machines would be a "general good" for labor was, he said, a "mistake." He had now decided—perhaps in response to the wrenching economic changes that the Industrial Revolution was causing at the time in Britain, Ricardo's home country—that these machines were, in fact, "often very injurious."[15]

This anxiety over the harmful impact of machines continued through the twentieth century. In the last few years, we have seen a frenzy of books and articles and reports on the threat of automation. Yet even as early as 1940, the debate about technological unemployment was so commonplace that the *New York Times* felt comfortable calling it an "old argument."[16] And it is true that these arguments do tend to repeat themselves. President Barack Obama, in his 2016 farewell address, described automation as "the next wave of economic dislocation." But so did President John F. Kennedy, about sixty years earlier, when, using almost identical words, he said that automation carried with it "the dark menace of industrial dislocation."[17] Similarly, in 2016 Stephen Hawking described how automation has "decimated" blue-collar work and predicted that this would soon "extend . . . deep into the middle classes."[18] Yet Albert Einstein had made a similar threat in 1931, warning that "man-made machines," which were meant to liberate human beings from drudgery and toil, were instead poised to "overwhelm" their creators.[19] In fact, in almost every decade since 1920, it is possible to find a piece in the *New York Times* engaging in some way with the threat of technological unemployment.[20]

UPHEAVAL AND CHANGE

Most of these anxieties about the economic harm caused by new technology have turned out to be misplaced. Looking back over the last few hundred years, there is little evidence to support the primary fear: that technological progress would create large pools of permanently unemployed workers. It is true that workers have been displaced by new technologies, but eventually most have also tended to find new work to do. Time and again, people have worried that "this time is different," that with the latest technologies mass displacement really is just around the corner—but, in fact, each time has been roughly the same, with mass displacement failing to emerge.

Understandably, this is a common cause for optimism among people who wonder what lies ahead. If those who worried in the past about the future of work were wrong to be concerned, then surely those who worry today are wrong to be anxious, too?

As we shall see, the issue is not so simple. Even if the "this time is different" worry was wrong before, it might still be right today. What's more, even if history were to repeat itself, we should still beware an excessively optimistic interpretation of the past. Yes, people did tend to find new work after being displaced by technology—but the way in which this happened was far from being gentle or benign. Take the Industrial Revolution again, that textbook moment of technological progress. Despite the Luddites' fears, the unemployment rate in Britain remained relatively low, as we can see in Figure 1.2. But, at the same time, whole industries were decimated, with lucrative crafts like hand weaving and candle making turned into profitless pastimes. Communities were hollowed out and entire cities thrust into decline. It is noteworthy that real wages in Britain barely rose—a measly 4 percent rise in total from 1760 to 1820. Meanwhile food became more expensive, diets were poorer, infant mortality worsened, and life expectancy fell.[21] People were, quite literally, diminished: a historian reports that average physical heights fell to their "lowest ever levels" on account of this hardship.[22]

Luddites are often dismissed today as technologically illiterate fools, but the evidence suggests they had legitimate grievances. Indeed, the

Figure 1.2: The Unemployment Rate in Britain, 1760–1900[23]

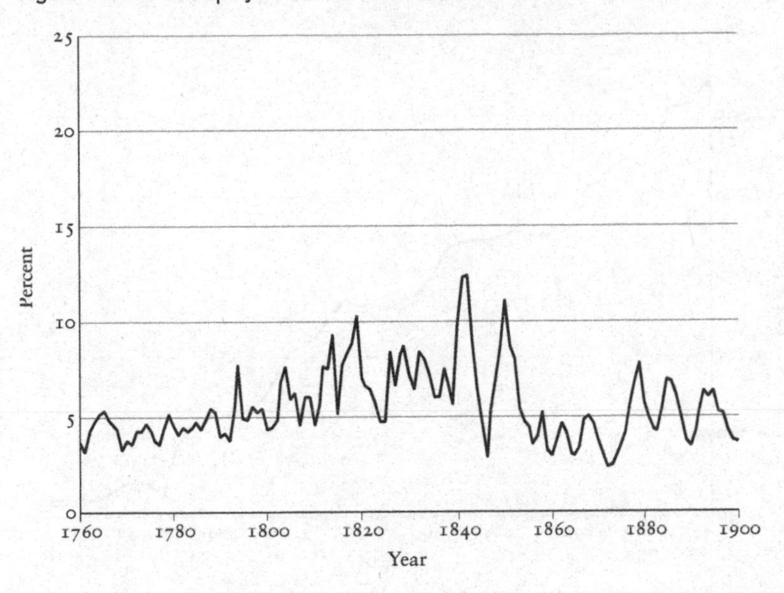

upheaval and distress caused by technological change eventually contributed to the case for the welfare state, perhaps the most radical invention of the twentieth century. None of what has been said about displaced workers eventually finding new jobs feels like cause for celebration. To paraphrase the economist Tyler Cowen, perhaps the future will be like the past—and that is why we ought *not* to be optimistic about the future of work.[24]

Nor is it the case, at a quick glance, that those who worried there might actually be less work in the future were completely wrong. Take Keynes who, in 1930, mused that within a hundred years technological progress would carry us into a world of "three-hour shifts" or a "fifteen-hour week." Today, his critics note with glee that his prediction will expire in a decade or so, and so far his "age of leisure" is not even on the horizon.[25] And there is some force to this criticism. But look a little deeper under the headline numbers, and the picture is more nuanced. In the OECD—the Organisation for Economic Cooperation and Development, a club of several dozen rich countries—the average number of hours that people work each year has continuously fallen over the past

Figure 1.3: Hours Worked per Person, per Year, in OECD Countries[26]

fifty years. The decline has been slow, about forty-five hours a decade, but steady nonetheless.

Importantly, a large part of this decline appears to be associated with technological progress and the increases in productivity that came along with it. Germany, for instance, is among the most productive countries in Europe, and also the one where people work the fewest hours a year. Greece is among the least productive, and—contrary to what many might think—the one where people work the most hours a year. As Figure 1.4 shows, this is a general trend: in more productive countries, people tend to work fewer hours. We may not be settling back into fifteen-hour weeks quite yet, as Keynes expected, but thanks to sustained technological progress we have begun to drift in that direction.[27]

All this is useful to have in mind when thinking about what might lie ahead. Today, we spend a lot of time trying to figure out the number of "jobs" that there will be in the future. Pessimists, for instance, imagine a world where lots of people find themselves idly sitting around, with nothing particularly productive to do, because the "robots" have

Figure 1.4: Productivity vs. Annual Hours Worked, 2014[28]

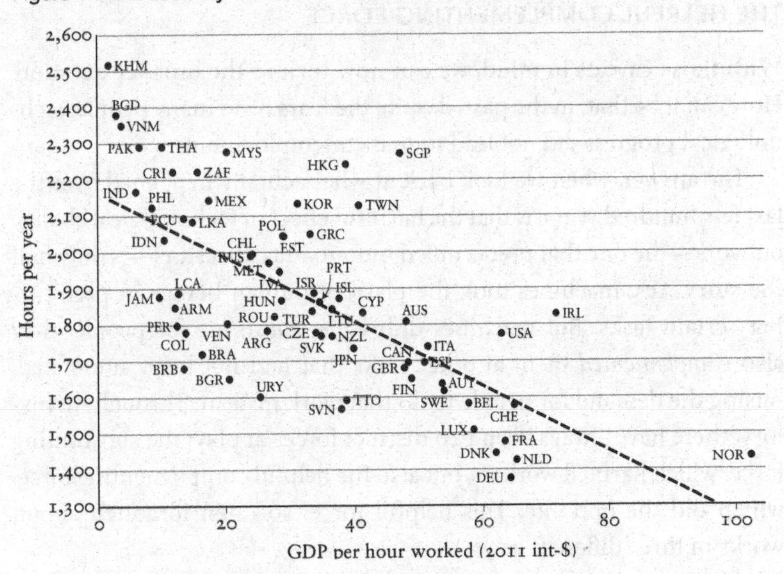

all the jobs. Optimists, in response, point to today's unemployment figures, at record lows in many places, and say that fears of a future with no jobs are baseless. But in this argument, both sides are thinking about the future of work in a very narrow way, as if all that matters is whether someone is employed or not. History suggests that this way of thinking, in terms of "jobs" alone, cannot capture the whole picture. Technological change may affect not only the amount of work, but also the *nature* of that work. How well-paid is the work? How secure is it? How long is the working day, or the working week? What sort of tasks does the work involve—is it the fulfilling sort of activity you leap out of bed in the morning to do, or the sort that keeps you hiding under the covers? The risk, in focusing on jobs alone, is not so much failing to see the proverbial forest for the trees, but failing to see all the different trees in the forest.

For now, I will continue to talk about "jobs." But we should bear in mind that technological progress will affect the world of work not only in the number of jobs there are for people to do, but in many other ways as well. In later chapters I will return to look at those more closely.

THE HELPFUL COMPLEMENTING FORCE

With those caveats in mind, we can now turn to the broader question. How can it be that, in the past, despite the fears of so many people, technological progress did not lead to mass unemployment?

The answer, when we look back at what actually happened over the last few hundred years, is that the harmful effect of technological change on work—the one that preoccupied our anxious ancestors—is only half the story. Yes, machines took the place of human beings in performing certain tasks. But machines didn't just *substitute* for people; they also *complemented* them at other tasks that had not been automated, raising the demand for people to do that work instead. Throughout history, there have always been two distinct forces at play: the substituting force, which harmed workers, but also the helpful complementing force, which did the opposite. This helpful force, so often forgotten about, works in three different ways.

The Productivity Effect

Perhaps the most obvious way that the complementing force helps human beings is that new technologies, even if they displace some workers, often make other workers more productive at their tasks. Think of the British weavers who were fortunate enough to find themselves operating one of Kay's flying shuttles in the 1730s, or one of Hargreaves's spinning jennies in the 1760s. They would have been able to spin far more cotton than their contemporaries who relied on their hands alone. This is the productivity effect.[29]

We can see this productivity effect at work today, too. Take a taxi driver who uses a sat-nav system to follow unfamiliar roads, an architect who uses computer-assisted design software to craft more complex buildings, or an accountant who uses tax computation software to handle harder, more intractable calculations. All are likely to become better at their tasks as a result. Or take doctors. In 2016, a team of researchers at MIT developed a system that can detect whether or not a breast biopsy is cancerous with 92.5 percent accuracy. Human pathologists, by comparison, were able to achieve a rate of 96.6 percent—but when they made their diagnoses with the MIT system alongside them, they were able to boost that rate to 99.5 percent, near perfection. The new technology made these doctors even better at the task of identifying cancers.[30]

In other settings, new technologies may automate some tasks, taking them out the hands of workers, but make those same workers more productive at the tasks that remain for them to do in their jobs. Think of a lawyer who is displaced from the task of looking through stacks of papers by an automated document review system, a piece of software that can scan legal material far more swiftly—and, in many cases, more precisely, too.[31] The same lawyer can now turn her attention to other tasks involved in providing legal advice, perhaps meeting face-to-face with her clients or applying her problem-solving skills to a particularly tricky legal conundrum.

In all these cases, if productivity increases are passed on to customers via lower prices or better-quality services, then the demand for whatever goods and services are being provided is likely to rise, and the demand for human workers along with it. Through the productivity effect, then, technological progress complements human beings in a very direct way, increasing the demand for their efforts by making them better at the work that they do.

The Bigger-Pie Effect

Economic history also reveals a second, less direct way that the complementing force has helped human workers: if we think again of the economy as a pie, technological progress has made the pie far bigger. As previously noted, over the last few hundred years, economic output has soared. The UK, for instance, has seen its economy grow 113-fold from 1700 to 2000. And that is nothing compared to other countries that were less developed at the start of this period: over the same three hundred years, the Japanese economy grew 171-fold, the Brazilian 1,699-fold, the Australian 2,300-fold, the Canadian 8,132-fold, and the US economy a whopping 15,241-fold.[32]

Intuitively, growth like this is likely to have helped workers. As an economy grows, and people become more prosperous with healthier incomes to spend, the opportunities for work are likely to improve. Yes, some tasks might be automated and lost to machines. But as the economy expands, and demand for goods and services rises along with it, demand will also increase for all the tasks that are needed to produce them. These may include activities that have not yet been automated, and so displaced workers can find work involving them instead.

Larry Summers, onetime director of the US president's National Economic Council, remembers making this point in his youth. In the 1970s, as a budding academic at MIT, he found himself tangled up in debates about automation. In his words, on campus at that time, "the stupid people thought that automation was going to make all the jobs go away" but "the smart people understood that when there was more produced, there would be more income, and therefore there would be more demand."[33] David Autor, perhaps today's most important labor market economist, makes a similar point, arguing that "people are unduly pessimistic . . . as people get wealthier, they tend to consume more, so that also creates demand."[34] And Kenneth Arrow, a Nobel Prize–winning economist, likewise argued that historically, "the replacement of men by machines" has not increased unemployment. "The economy does find other jobs for workers. When wealth is created, people spend their money on something."[35]

The Changing-Pie Effect

Finally, the last few hundred years also suggest a third way for the complementing force to work. Thanks to technological progress, economies have not only grown, they have also transformed—producing very different output, in very different ways, at different moments in history. If we think again of the economy as a pie, new technologies have not only made the pie bigger, but *changed the pie*, too. Take the British economy, for example. Its output, as we noted, is now more than a hundred times what it was three centuries ago. But that output, and the way it is produced, has also completely transformed. Five hundred years ago, the economy was largely made up of farms; three hundred years ago, of factories; today, of offices.[36]

Again, it is intuitive to see how these changes might have helped displaced workers. At a certain moment, some tasks might be automated and lost to machines. But as the economy changes over time, demand will rise for other tasks elsewhere in the economy. And since some of these newly in-demand activities may, again, not have been automated, workers can find jobs involving them instead.

To see this changing-pie effect in action, think about the United States. Here you can see displaced workers tumbling through a changing economy, time and again, into different industries and onto different tasks. A century ago, agriculture was a critical part of the American economy: back in 1900, it employed two in every five workers. But since then, agriculture has collapsed in importance and today it employs fewer than two in every *hundred* workers.[37] Where did the rest of those workers go? Into manufacturing. Fifty years ago, that sector superseded agriculture: in fact, in 1970, manufacturing employed a quarter of all American workers. But then that sector also went into relative decline and today fewer than a tenth of American workers are employed in it.[38] Where did these displaced factory workers go? The answer is the service sector, which now employs more than eight in ten workers.[39] And there is nothing distinctly American about this story of economic transformation, either. Almost all developed economies have followed a similar path, and many less-developed economies are following it, too.[40] In 1962, 82 percent of Chinese workers were employed in agriculture; today, that has fallen to around 31 percent, a larger and faster decline than the American one.[41]

Where the "bigger pie" effect suggests that our anxious predecessors were shortsighted, unable to see that economies would grow in the future, this "changing pie" effect suggests that they also suffered from a failure of imagination. Our ancestors could not see that, in years to come, what their economies produced and how they produced it would transform beyond recognition. To an extent, this failure is understandable. In 1900, for instance, most English people worked on farms or in factories. Few could have anticipated that in the future, a single "health care" organization, the National Health Service, would employ far more people than the number of men then working on all the farms in the country combined.[42] The health care industry itself did not exist back then in the way we think about it now, and the idea that this massive health care employer would be the British government would have seemed even more bizarre; at the time, after all, most health care was privately or voluntarily provided. The same thing is true for so many roles today, too: job titles like search engine optimizer, cloud computing specialist, digital marketing consultant, and mobile app developer would have been impossible to envision even a few decades ago.[43]

THE BIG PICTURE

The idea that the effect of technology on work might depend upon the interaction between these two rival forces—a harmful substituting force and a helpful complementing force—is not new. However, these forces tend not to be explained in a particularly clear way. Books, articles, and reports on automation can be confusing, hinting at these two effects but often using wildly different terms. Technology, they say, displaces and augments, replaces and enhances, devalues and empowers, disrupts and sustains, destroys and creates. The challenge is to compete with computers and to cooperate with them, to race against the machines and run alongside them. There is talk of the rise of machines and the advance of humans, of threatening robots and comforting co-bots, of the artificial intelligence of machines and the augmented intelligence of human beings. The future, they say, holds both obsolescence and ever-greater relevance; technology is a threat and an opportunity, a rival and a partner, a foe and a friend.

The discussion of economic history in this chapter, brief though it may be, should clarify how these two forces actually work. On the one hand, a machine "substitutes" for human beings when it displaces them from particular tasks. This, when it happens, is relatively easy to see. On the other hand, a machine "complements" human beings when it raises the demand for their work at other tasks—a phenomenon that, as we have seen, can happen in three different ways, and is often less easy to identify than its destructive cousin.

Distinguishing clearly between the substituting and complementing effects of technology helps to explain why past anxieties about technological unemployment were repeatedly misplaced. In the clash between these two fundamental forces, our ancestors tended to pick the wrong winner. Time and again, they either neglected the complementing force altogether, or mistakenly imagined that it would be overwhelmed by the substituting force. As David Autor puts it, people tended "to overstate the extent of machine substitution for human labor and ignore the strong complementarities between automation and labor."[44] As a result, they repeatedly underestimated the demand for the work of human beings that would remain. There was always, by and large, enough to keep people in employment.

We can see this playing out with individual technologies, too. Consider, for instance, the story of the automatic teller machine. When the ATM was invented, it was designed to displace bank tellers from the task of handing over cash. It was part of the self-service culture that spread through economic life in the mid-twentieth century, along with do-it-yourself gas stations, check-yourself-out cash registers, serve-yourself candy dispensers, and so on.[45] The first ATM is said to have been installed in Japan in the mid-1960s.[46] The machines became popular in Europe a few years later—a solution, in part, to the problem of increasingly powerful unions that demanded banks close on Saturday, the only day many working customers could visit. In the United States, the number of ATMs more than quadrupled from the late 1980s to 2010, by which point there were more than four hundred thousand of them in action. Given that sort of uptake, you might have expected a precipitous fall in the number of tellers employed in American banks. But instead the opposite happened: the number of tellers also rose during that period, by as much as 20 percent.[47] How can we explain that puzzle?

We can use the two forces we've discussed to get a sense of what happened. The answer is that ATMs did not simply substitute for bank tellers, but also complemented them. Sometimes they did so directly: ATMs didn't make tellers more productive at handing out cash, but they did free them up to focus their efforts on other activities, like offering face-to-face support and providing financial guidance. This meant better service for those who walked into a branch, attracting more customers. ATMs also helped to reduce the cost of running branches, allowing banks to draw even more footfall by offering better prices.

At the same time, ATMs complemented tellers indirectly. Partly, this may have been the bigger-pie effect: as ATMs and countless other innovations boosted the economy over those years, incomes rose, and so did demand for banks and the remaining tellers that worked in them. And partly, it may have been the changing-pie effect as well: as people became more prosperous, their demand may have swung away from simply depositing and withdrawing money, and more toward the "relationship-banking" services that tellers were now providing.

Together, all these helpful effects meant that although the number of tellers needed at an average branch fell from twenty in 1988 to thirteen in 2004, the number of branches rose during that time—in urban areas by as much as 43 percent—to meet the growing demand for banking services. This meant more work for bank tellers overall, and that is why the number of tellers rose rather than fell.[48]

Of course, the full history of work and technology is more complex and nuanced than that set out in this chapter. The story has not always been as clear at different times and in different places. But these are the general contours. Technological progress has brought many disruptions and dislocations, as we have seen; but from the Industrial Revolution until today, workers who worried that machines would permanently replace them have largely been proven wrong. Up until now, in the battle between the harmful substituting force and the helpful complementing force, the latter has won out, and there has always been a large enough demand for the work that human beings do. We can call this the Age of Labor.

2

The Age of Labor

The Age of Labor can be defined as a time when successive waves of technological progress have broadly benefited rather than harmed workers. But although this progress has been good for workers in general, not all of them have always benefited. Nor have the benefits been consistent over time: technological progress has proven to be a fickle friend, with different groups of workers gaining more from it at different moments. To make sense of these developments, in the last decade or two, many economists have had to radically change the story they tell about technology and its impact on work.

We might not think of economists as storytellers, but that is what they are. Their stories just happen to be written in a foreign language, mathematics, in an attempt to make their narrative precise for fluent readers (but making them frustratingly unintelligible for those who are not). They are meant to be nonfiction, rooted in the facts, the plot aligning as closely as possible with reality. Some are epics, trying to capture great swaths of human activity in a single heroic sweep; others are far more limited, tightly focused on explaining very particular patterns of behavior. Economists prefer to call them "models" rather than stories, which certainly sounds like a weightier label. But in the end, any model is simply a tale told in equations and charts, designed to capture an insight about how the real world works.

THE TWENTIETH CENTURY AND BEFORE

For most of the second half of the twentieth century, the workers who appeared to benefit the most from technological change were those who had more years of formal schooling behind them. And economists developed a story to explain why that was so, which went something like this.[1]

The main character in this story is the digital electronic computer. Invented around the middle of the twentieth century, it grew explosively in power and usefulness as time went on. In the late 1950s and early 1960s, businesses started to make extensive use of mainframe computers.[2] Then the personal computer (PC) was invented and began to spread; as late as 1980, the United States had fewer than one PC per hundred people, but by the turn of the century that figure had risen to more than sixty.[3] What's more, these machines became far more capable over time. The number of computations that a machine could perform in a given amount of time soared throughout the second half of the century.[4] This is shown in Figure 2.1, which begins with computing manually by hand in 1850, and ends with a Dell Precision Workstation 420 desktop computer in 2000, with a spread of other machines in between.

To capture in a manageably sized chart quite how fast these computations per second rose, the vertical axis here has a logarithmic scale. This means that as you move up the vertical axis, each step represents a tenfold increase in computations per second (two steps is a hundredfold increase, three steps a thousandfold, and so on). As we can see, just from 1950 to 2000 computational power increased roughly by a factor of ten billion.

But while these powerful new machines might have been able to handle certain useful workplace tasks, like performing complex numerical calculations or typesetting text in an attractive way, they did not do away with the demand for the work of human beings altogether. In fact, these computers led to far greater demand for the sorts of high-skilled people who were able to operate them and put them to productive use. Other technologies that were emerging at the time are thought to have had the same effect, creating a demand for high-skilled workers who

Figure 2.1: Index of Computations per Second, 1850–2000 (manual calculation = 1)[5]

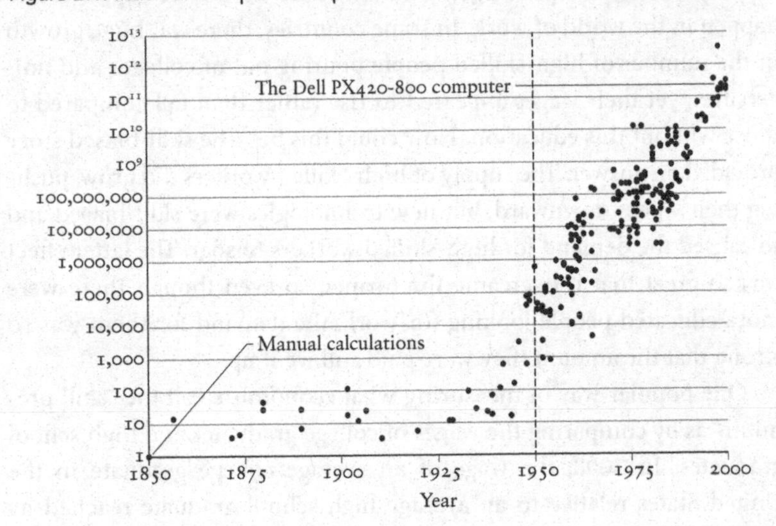

were capable of using them effectively. In this account, then, technological change did not benefit all workers equally, but had a particular tilt. As economists put it, it was "skill-biased."

(It is worth noting that in this story, economists use a very particular definition of what it means to be "skilled": namely, the amount of formal schooling someone has under their belt. This definition can puzzle non-economists, since many people we might think of as "skilled" in a more everyday use of the term—an expert hairdresser or a nimble-fingered gardener—are often deemed "unskilled" by economists because they have not gone to college. There is, in short, a divergence here between the commonsense and the "economist-sense" use of the word *skilled*. This does not mean either usage is wrong. But it does mean that to avoid confusion and offense, it is important to be clear what exactly economists are talking about when they coin terms like "skill-biased.")

This skill-biased story of technological progress in the second half of the twentieth century was strongly supported by the evidence, and it neatly explained an empirical puzzle that emerged around that time. A basic principle in economics is that when the supply of something goes up, its price should go down. The puzzle was that in the twentieth century, there were prolonged periods where the reverse appeared to happen in the world of work. In some countries, there was huge growth in the number of high-skilled people pouring out of colleges and universities, yet their wages appeared to rise rather than fall compared to those without this education. How could this be? The skill-biased story provided an answer. The supply of high-skilled workers did grow, pushing their wages downward, but new technologies were skill-biased and so caused the demand for high-skilled workers to soar. The latter effect was so great that it overcame the former, so even though there were more educated people looking for work, the demand for them was so strong that the amount they were paid still went up.

One popular way of measuring what economists call the "skill premium" is by comparing the wages of college graduates and high school graduates. In 2008, the wage of an average college graduate in the United States relative to an average high school graduate reached its highest level in decades, as shown in Figure 2.2. (The comparison is plotted there as the "log wage gap," the logarithm of the ratio of the average wages of the two groups; the 2008 log wage gap was 0.68, which

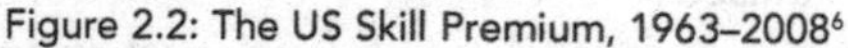
Figure 2.2: The US Skill Premium, 1963–2008[6]

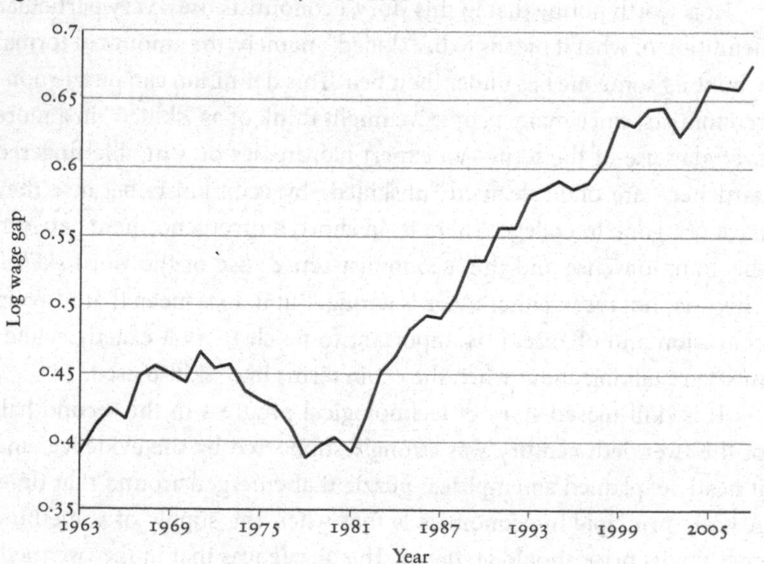

implies that the earnings of an average college graduate were almost double the earnings of an average high school graduate.)[7] Other countries followed very similar patterns in that period as well.[8]

Another way to see the skill-biased story at work is to look at how wages have changed over time for a variety of different levels of schooling. This is shown in Figure 2.3. As the charts show, people with more years of schooling not only tend to earn more at every point in the past half century, but the gap between them and those with less schooling has tended to grow over time as well. (For women, this story becomes clearer from the 1980s onward.)

However, while the skill-biased story does a good job of explaining what happened to the world of work in the latter part of the twentieth century, before that time the picture was very different. Consider Figure 2.4, which shows the skill premium in England dating back to 1220. (Luckily, there is indeed data that stretches that far back: English institutions have proven to be both remarkably stable and uncommonly assiduous in their record keeping for the past millennium.) Given there were few college degrees back in 1220, the skill premium here is instead measured by comparing the wages of craftsmen to those of laborers. And

Figure 2.3: Real Wages of Full-Time US Workers, 1963–2008 (Index 1963 = 100)[9]

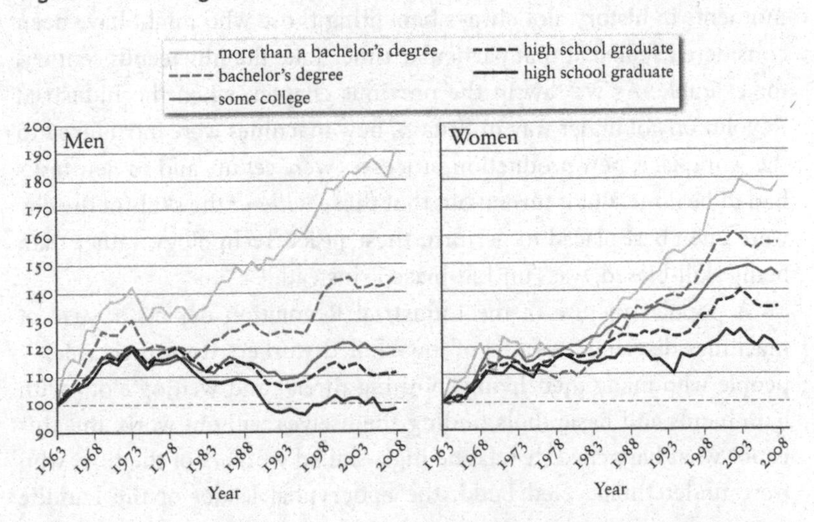

Figure 2.4: The English Skill Premium, 1220–2000[10]

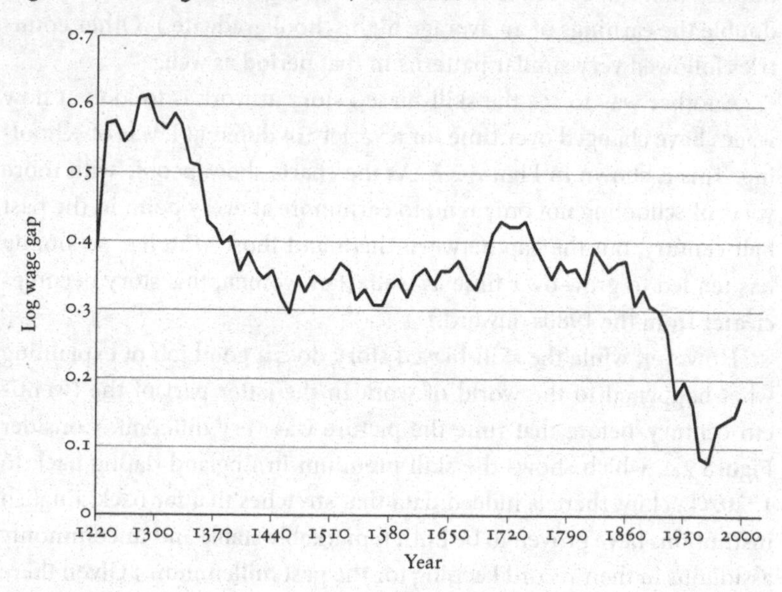

as we can see, over the long run this skill premium no longer follows the same upward pattern that we saw in Figure 2.3.

What is going on here? This longer view suggests that technological change has in fact favored different types of workers at different moments in history, not always benefiting those who might have been considered skilled at that particular time. Take the nineteenth century, for example. As we saw in the previous chapter, when the Industrial Revolution got under way in Britain, new machines were introduced to the workplace, new production processes were set up, and so new tasks had to be done. But it turned out that those *without* the skills of the day were often best placed to perform these tasks. Technology, rather than being skill-biased, was "unskill-biased" instead.[11]

A popular picture of the Industrial Revolution depicts a wave of machines displacing swaths of low-skilled workers from their roles—people who made their living spinning thread and wefting cloth with bare hands and basic tools finding themselves without work. But this is not what happened. It was the high-skilled workers of the time who were under threat. Ned Ludd, the apocryphal leader of the Luddite uprising against automation, was a skilled worker of his age, not an unskilled one. If he actually existed, he would have been a professional of sorts—perhaps even a card-carrying member of the Worshipful Company of Clothworkers, a prestigious club for people of his trade. And the mechanical looms that displaced Ned and his comrades meant that someone with less skill, without Ned's specialized training, could take his place. These new machines were "de-skilling," making it easier for less-skilled people to produce high-quality wares that would have required skilled workers in the past.

The share of unskilled workers in England appears to have doubled from the late 1500s to the early 1800s.[12] This change was no accident. Andrew Ure, an influential figure who acted as a sort of early management consultant to manufacturers, called for taking away tasks from "the cunning workman" and replacing him with machines so simple to use that "a child may superintend" instead. (He did not mean this metaphorically: child labor was an acceptable practice back then.)[13] And as the economic historian Joel Mokyr notes, this trend was not confined to the world of cotton and cloth: "First in firearms, then in clocks, pumps, locks, mechanical reapers, typewriters, sewing machines, and eventually in engines and bicycles, interchangeable parts technology proved superior and replaced the skilled artisans working with chisel and file."[14]

At the turn of the twenty-first century, then, the conventional wisdom among economists was that technological progress was sometimes skill-biased, at other times unskill-biased. In either case, though, many economists tended to imagine that this progress always broadly benefited workers. Indeed, in the dominant model used in the field, it was *impossible* for new technologies to make either skilled or unskilled workers worse off; technological progress always raised everyone's wages, though at a given time some more than others. This story was so widely told that leading economists referred to it as the "canonical model."[15]

A NEW STORY IN THE TWENTY-FIRST CENTURY

The canonical model dominated discussion among economists for decades. But recently, something very peculiar began to happen. Starting in the 1980s, new technologies appeared to help both low-skilled

Figure 2.5: Percentage Point Change in Share of Total Employment, 1995–2015[16]

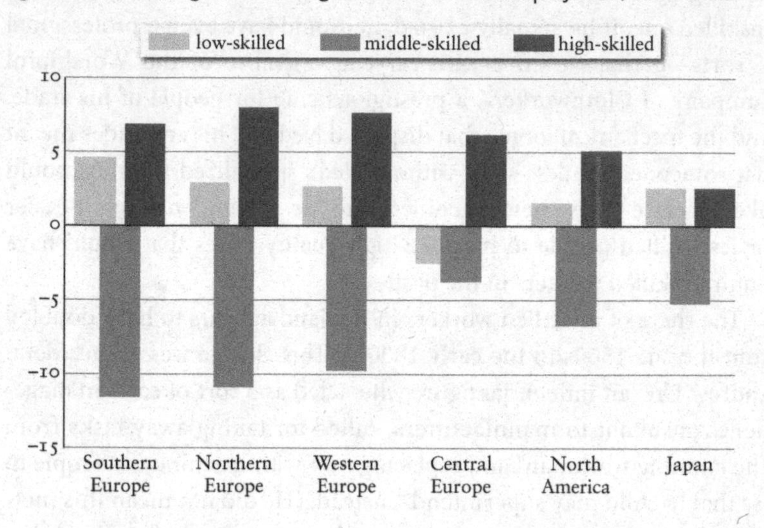

and high-skilled workers at the same time—but workers with middling skills did not appear to benefit at all. In many economies, if you took all the occupations and arranged them in a long line from the lowest-skilled to the highest-skilled, over the last few decades you would have often seen the pay and the share of jobs (as a proportion of total employment) grow for those at either end of the line, but wither for those near the middle. We can clearly see this trend in Figure 2.5.

This phenomenon is known as "polarization" or "hollowing out." The traditionally plump midriffs of many economies, which have provided middle-class people with well-paid jobs in the past, are disappearing. In many countries, as a share of overall employment there are now more high-paid professionals and managers—as well as more low-paid caregivers and cleaners, teacher's aides and hospital assistants, janitors and gardeners, waiters and hairdressers.[17] But there are fewer middling-pay secretaries and administrative clerks, production workers and salespeople.[18] Labor markets are becoming increasingly two-tiered and divided. What's more, one of these tiers is benefiting far more than the other. The wages of people standing at the top end of the lineup, the 0.01 percent who earn the most—in the United States, the 16,500 families with an income over $11,300,000 per year—have soared over the last few decades.[19]

(A terminological point, again: this presentation of the data may seem like it is treating "pay" and "skill" as if they are the same thing—as if lining up workers from the lowest-paid to the highest-paid is the same exercise as lining them up from low-skilled to high-skilled. As before, the issue has to do with economists' idiosyncratic definition of the word *skilled*. Clearly, there exist jobs that are low-paid but require significant skills in the common sense of the word—emergency medical technicians, for example. And there are also jobs that are high-paid but, many believe, require very little skill at all—recall, for instance, the caustic complaints made after the 2007–8 economic crisis about those who work in finance. But, as noted earlier, when economists talk about "skill," they really mean "level of formal schooling." And using pay as a proxy measure for that particular definition of skill does turn out to be a reasonable thing to do: as we have seen, people with more schooling behind them tend to earn more as well. So whether you line up jobs according to their level of pay or the average number of years of schooling that people in them have does not really matter—the hollowing-out pattern looks roughly the same.[20])

The hollowing out of the labor market was a new puzzle. And the canonical model that dominated economic thinking in the late twentieth century was powerless to solve it. It was narrowly focused on just two groups of workers, the low-skilled and the high-skilled, and had no way to explain why middling-skilled workers were facing such a very different fate from their low- and high-skilled contemporaries. A new account was needed. Economists went back to their intellectual drawing boards. And over the past decade or so, intellectual support has emerged for an entirely different way of thinking about technology and work. Pioneered by a group of MIT economists—David Autor, Frank Levy, and Richard Murnane—it became known as the "Autor-Levy-Murnane hypothesis," or the "ALM hypothesis" for short.[21] A decade ago, when I began to think seriously about the future, this was the story I was handed to help me do so.[22]

The ALM hypothesis built upon two realizations. The first of these was simple: looking at the labor market in terms of "jobs," as we often do, is misleading. When we talk about the future of work, we tend to think in terms of journalists and doctors, teachers and nurses, farmers and accountants; and we ask whether, one day, people who have one of these jobs might wake up and find a machine in their place. But thinking like this is unhelpful because it encourages us to imagine that a given job is a uniform, indivisible blob of activity: lawyers do "lawyering," doctors "doctoring," and so on. If you look closely at any particular job, though, it is obvious that people perform a wide variety of different *tasks* during their workday. To think clearly about technology and work, therefore, we have to start from the bottom up, focusing on the particular tasks that people do, rather than looking from the top down, looking only at the far more general job titles.

The second realization was subtler. With time, it became clear that the level of education required by human beings to perform a given task—how "skilled" those people were—was not always a helpful indication of whether a machine would find that same task easy or difficult. Instead, what appeared to matter was whether the task itself was what the economists called "routine." By "routine," they did not mean that the task was necessarily boring or dull. Rather, a task was regarded as "routine" if human beings found it straightforward to explain how they performed it—if it relied on what is known as "explicit" knowledge, knowledge which is easy to articulate, rather than "tacit" knowledge, which is not.[23]

Autor and his colleagues believed that these "routine" tasks must be easier to automate. Why? Because when these economists were trying to determine which tasks machines could do, they imagined that the only way to automate a task was to sit down with a human being, get her to explain how she would perform that task, and then write a set of instructions based on that explanation for machines to follow.[24] For a machine to accomplish a task, Autor wrote, "a programmer must first fully understand the sequence of steps required to perform that task, and then must write a program that, in effect, causes the machine to precisely simulate these steps." If a task was "non-routine"—in other words, if human beings struggled to explain how they performed it—then it would be difficult for programmers to specify it as a set of instructions for the machine.[25]

The ALM hypothesis brought these two ideas together. Machines, it said, could readily perform the "routine" tasks in a job, but would struggle with the "non-routine" tasks. This clever argument could explain the strange trends captured in Figure 2.5—since when economists broke down an assortment of different jobs into their constituent tasks, many of the activities that middling-paid people performed in their work turned out to be "routine," whereas those done by the low-paid and high-paid were not. That was why labor markets around the world were being hollowed out, taking on hourglass figures. Technological change was eating away at the "routine" tasks clustered in the middle, but the "non-routine" tasks at either end were indigestible, left for human beings to undertake.[26]

That high-paid, high-skilled work often turned out to be "non-routine" was not surprising. The tasks involved in those jobs required human faculties like creativity and judgment, which are very hard or outright impossible to capture in a set of rules. (Most people would be very suspicious of any definitive set of instructions for "how to be creative," for example.) But why did low-paid, low-skilled work also turn out to be "non-routine"? The explanation is partly that this work was often part of the service economy, and the interpersonal skills required to provide services were hard to capture in rules. But it was also because low-paid work often required manual skills that were hard to automate. Computer scientists were already familiar with this finding: many of the basic things we do with our hands are the most difficult tasks for a machine to do. (This is known as "Moravec's Paradox," after Hans Moravec, a futurist and inventor who was one of the first to note it down.[27]) When human beings perform tasks like cooking a meal or trimming a shrub, they tend to perform them unconsciously and instinctively, without deliberate thought. Therefore, while people might find these tasks simple to do, they may also find it very difficult to explain *how* to do them. Accordingly, it seemed, tasks like these could not be readily automated, either.

Technological progress, it appeared, was neither skill-biased nor unskill-biased, as the old stories had implied. Rather it was *task-biased*, with machines able to perform certain types of tasks but unable to perform others. This meant that the only workers to benefit from technological change would be those well placed to perform the "non-routine" tasks that machines could not handle. In turn, this explained why certain types of middling-skilled workers might not gain from new technology at all—if they found themselves stuck in jobs made up largely of "routine" tasks that machines could handle with ease.

INSIGHTS FROM THE ALM HYPOTHESIS

It may be surprising to hear that economists were so willing to change their minds, swinging from a view of technology as always benefiting workers to a task-biased view instead. There is an old story on this theme, though: Keynes, when pressed by a critic for shifting his position on some economic issue, supposedly responded, "When the facts change, I change my mind. What do you do, Sir?"[28] This is often quoted in jest, a witty example of how to concede a mistake and evade it at the same time. But remember what economists actually do: they tell stories, mathematical tales meant to capture reality. So this is exactly how we ought to want economists to behave: to adjust when the facts change, to update their models, and to re-craft their stories. And that is exactly what economists who think about the labor market have done in recent decades. Far from being a sign of intellectual inconsistency, this is a good thing.

The ALM hypothesis also helps to expose several types of mistaken thinking about the future of work. For instance, it is very common to hear discussions about the chances of various jobs being automated, with statements like "nurses are safe but accountants are in trouble" or "X percent of jobs in the United States are at risk from automation but only Y percent in the UK." One influential study, by Oxford's Carl Frey and Michael Osborne, is often reported as claiming that 47 percent of US jobs are at risk of automation in the coming decades, with telemarketers the most at risk (a "99 percent" risk of automation) and recreational therapists the least (a "0.2 percent" risk).[29] But as Frey and Osborne themselves have noted, conclusions like this are very misleading. Technological progress does not destroy entire jobs—and the ALM "job" versus "task" distinction explains why. No job is an unchanging blob of activity that can be entirely automated in the future. Rather, every job is made up of many tasks, and some of these tasks are far easier to automate than others. It is also important to remember that as time passes, the tasks that make up a particular occupation are likely to change. (There are few jobs, if any, that look the same today as they did thirty years ago.)

The point is driven home by a 2017 study carried out by McKinsey & Company, which reviewed 820 different occupations. Fewer than 5 percent of these, they found, could be completely automated with existing technologies. On the other hand, more than 60 percent of the occupations were made up of tasks of which at least 30 percent could be automated.[30] In other words, very few jobs could be entirely done by machines, but most could have machines take over at least a significant part of them.

That's why those who claim that "my job is protected from automation because I do X," where "X" is a task that is particularly difficult to automate, are falling into a trap. Again, no job is made up of one task: lawyers do not only make court appearances, surgeons do not only perform operations, journalists do not only write original opinion pieces. Those particular tasks might be hard to automate, but that does not necessarily apply to all of the other activities these same professionals do in their jobs. Lawyers, for instance, may argue that no machine could stand and deliver a stunning peroration to a gripped jury—and they may well be right about that. But machines today certainly *can* retrieve, assemble, and review a wide range of legal documents, tasks that make up a big part of most lawyers' jobs—and, in the case of junior lawyers, almost their entire jobs.

Technological optimists make a similar mistake when they point out that, of 271 occupations in the 1950 US Census, only a single one—elevator operators—has subsequently disappeared due to automation.[31] This is not a sign of technology's impotence, as they might imagine. It is further evidence that the important changes are deeper, taking place at the level of the underlying tasks rather than at the level of the job titles themselves.

The second key realization behind the work of Autor and his colleagues—that what matters is the nature of the tasks themselves, not whether the worker performing them is "skilled" or not—is also a crucial insight. White-collar professionals are often startled by it, given the time and money they have devoted to their education. Some even take offense, imagining that a rude equivalence is being drawn between the "sophisticated" work that they do and the unrefined labor of others. But the point is that their work is not as special as they imagine. Once you break down most professional jobs into the tasks that make them up, many of these tasks turn out to be "routine" and can already be automated. The fact that educated professionals tend to use their heads, rather than their hands, to perform their task does not matter. Far more important is whether the tasks are "routine."

AN OPTIMISTIC WAY OF THINKING

The ALM hypothesis is important not only because of its success in explaining the economic peculiarities of the recent past—the hollowing out of the labor market and the harm caused to workers caught in the middle—but also because it explains the optimism that many forecasters feel about technology and the future.

The old "canonical model" of technological change also suggested an optimistic view of the future of work, but for a wildly unrealistic reason: in that model, as we saw, technology always complements workers (albeit some more than others). Today, few people would make an argument like that. Instead, those who are optimistic about the future of work build a case that looks more like the task-biased story of the ALM hypothesis. They argue that new technologies do substitute for workers, but not at everything, and that machines tend to increase the demand for human beings to perform tasks that cannot be automated. Autor himself captured this case for feeling optimistic in a pithy line: "tasks that cannot be substituted by automation are generally complemented by it."[32]

Arguments like this rely upon the assumption that there are some tasks that machines simply cannot do, and therefore there is a firm limit to the harmful substituting force. Of course, some might say that this proposition is intuitively obvious. But the ALM hypothesis provides formal reasoning to back up this intuition: machines cannot be taught to perform "non-routine" tasks, because people struggle to explain how they perform them. As Autor puts it, "the scope for this kind of substitution is bounded because there are many tasks that people understand tacitly and accomplish effortlessly but for which neither computer programmers nor anyone else can enunciate the explicit 'rules' or 'procedures.'"[33] So while future technologies may increasingly substitute for human beings in "routine" tasks, they will always complement human beings in the "non-routine" tasks that remain.

This distinction between "routine" and "non-routine" tasks has now spread far beyond academic economic papers. The most influential institutes and think tanks—from the IMF to the World Bank, from the OECD to the International Labour Organization—have relied on it to decide which human endeavors are at risk of automation.[34] Mark Carney, the governor of the Bank of England, has echoed it in a warning of a "massacre of the Dilberts": new technologies, he believes, threaten "routine cognitive jobs" like the one that employs Dilbert, the cubicle-bound comic strip character.[35] President Obama similarly warned that roles "that are repeatable" are at particular risk of automation.[36] And large companies have structured their thinking around the idea: the investment bank UBS claims that new technologies will "free people from routine work and so empower them to concentrate on more creative, value-added services"; the professional services firm PwC says that "by replacing workers doing routine, methodical tasks, machines can amplify the comparative advantage of those workers with problem-solving, leadership, EQ, empathy, and creativity skills"; and Deloitte, another professional services firm, reports that in the UK "routine jobs at high risk of automation have declined but have been more than made up for by the creation of lower-risk, non-routine jobs."[37]

Magazine writers and commentators have also popularized the concept. The *Economist*, for instance, explains that "what determines vulnerability to automation, experts say, is not so much whether the work concerned is manual or white-collar but whether or not it is routine." The *New Yorker*, meanwhile, asks us to "imagine a matrix with two axes, manual versus cognitive and routine versus non-routine," where every task is sorted into one of the quadrants.[38] And elsewhere we can see the shadow of the "routine" versus "non-routine" distinction in the way that people so often describe automation. Machines, they say, can only do things that are "repetitive" or "predictable," "rules-based" or "well-defined" (in other words, "routine" tasks); they cannot handle those that are "difficult to specify" or "complex" (the "non-routine" ones).

In fact, very few approaches in modern economic thought have been as influential as the ALM hypothesis. A set of ideas that began in the stillness of an economist's study have trickled down to the wider world, shaping the way in which many people think about what lies ahead.[39] The ALM hypothesis has encouraged us to believe that there is a wide range of tasks that can never be automated, a refuge of activity that will always provide enough work for human beings to do. The Age of Labor, to which we have become accustomed, will carry on.

In my opinion, this optimistic assumption is likely to be wrong. But to understand why, we must first take a look at the changes that have taken place in the world of technology and artificial intelligence.

3

The Pragmatist Revolution

Human beings have long shared stories about machines that could do remarkable things. Three thousand years ago, Homer told a tale of "driverless" three-legged stools, made by a god, that would scuttle toward their owner at his command.[1] Plato wrote of Daedalus, a sculptor so talented that his statues had to be tied up to stop them from running away.[2] This story, fanciful as it may seem, troubled Plato's student Aristotle so much that he wondered what would happen to the world of work if "every tool we had could perform its task, either at our bidding or itself perceiving the need."[3] The old Jewish sages wrote of mystical creatures called golems, fashioned out of mud and clay, which would come to life to help their owners at the muttering of the right incantation. One golem, called Yosef, is said to still lie hidden in the attic of the grand synagogue in Prague; centuries ago, according to legend, Rabbi Judah Loew brought him to life to protect the Jews in his community from persecution.[4]

Tales of this kind are scattered through ancient writing. But more recent history is full of them, too, fables of wonderful and strange machines that go about their work without any apparent human involvement. Today we call them "robots," but before that word was invented in 1920 they were called "automata"—and they were wildly popular. In the fifteenth century, Leonardo Da Vinci sketched out an autonomous cart and an armor-clad humanoid robot; he also designed a mechanical lion for the king of France, which, when whipped three times by his majesty, would open its chest to reveal the emblem of the monarchy.[5] In the eighteenth century, a Frenchman called Jacques de Vaucanson became famous for his machines: one that could play the flute, another that could clatter tambourines in time, and—his most celebrated—a duck that could eat, drink, flap its wings, and defecate. Disappointingly, the so-called *canard digérateur*, or "digesting duck," was not actually true to its name; a hidden compartment simply released a convincing alternative to the real thing (bread crumbs, dyed green).[6] Swindles like this were amusingly frequent. Around the same time, a Hungarian called Wolfgang von Kempelen built a chess-playing machine, nicknamed "The Turk" on account of its Eastern appearance. It toured the world for decades, defeating celebrated opponents like Napoleon Bonaparte and Benjamin Franklin. Little did they know that a human chess master was hidden in the belly of the supposed automaton.[7]

Why were people so captivated by these machines? In part, it was probably the spectacle: that some of them did amusing and (as with the digesting duck) often moderately offensive things. But what about the chess machine? Why did that excite everyone? It was not because of its manual dexterity: the eighteenth-century world was full of technologies that carried out tasks requiring physical capabilities far more impressive than moving a chess piece. Rather, people were impressed by the apparent ability of the Turk to carry out tasks that require cognitive capabilities, the sorts of things that human beings do with their heads rather than hands. Everyone would have imagined such activities to be far out of reach of any machine, and yet the chess-playing device did much more than move pieces aimlessly around the board. It seemed to ponder possible moves and outsmarted competent human players. It seemed to deliberate and reflect. In human beings, we recognize these capabilities as requiring "intelligence." And this is what would have shocked its audience: the machines appeared to act intelligently, too.

In the end, almost all these accounts were fictional. Many designs scribbled down by inventors remained speculative and were never built, and the machines that did get made tended to rely upon trickery. It is no coincidence that Jean-Eugène Robert-Houdin, the first modern magician (from whom the illusionist Houdini would take his name a few decades later), was also a master craftsman of automata, at one point getting called upon to repair the famous digesting duck itself when it broke a wing.[8] But in the twentieth century, all this changed. For the first time, researchers began to build machines with the serious intention of rivaling human beings—a proper, sophisticated program of constructing intelligence was under way. Their aspirations were now serious, no longer confined to fiction or dependent on deceit.

THE FIRST WAVE OF AI

At a 1947 meeting of the London Mathematical Society, Alan Turing told the gathering that he had conceived of a computing machine that could exhibit intelligence.[9] Turing deserved to be taken seriously: perhaps Britain's leading World War II code breaker, he is one of the greatest computer scientists to have ever lived. Yet the response to the ideas in his lecture was so hostile that within a year he felt compelled to publish a new paper on the topic, responding in furious detail to assorted objections to his claim that machines "could show intelligent behaviour." "It is usually assumed without argument it is not possible," he thundered in the opening lines. Turing thought that the objections were often "purely emotional"—"an unwillingness to admit the possibility that mankind can have any rivals in intellectual power," for instance, or "a religious belief that any attempt to construct such machines is a sort of Promethean irreverence."[10]

Less than a decade later, a group of four American academics—John McCarthy, Marvin Minsky, Nathaniel Rochester, and Claude Shannon—sent a proposal to the Rockefeller Foundation, asking for enough money to support a "2 month, 10 man study of artificial intelligence" at Dartmouth College. (The term "artificial intelligence" was McCarthy's invention.)[11] Their proposal was striking for its ambition and optimism. "Every aspect of learning or any other feature of intelligence," they claimed, could be simulated by a machine. And they believed that "a significant advance" could be made "if a carefully selected group of scientists work on it together for a summer."[12]

As it happened, no particular advance worth celebrating was made at

Dartmouth that summer of 1956. Nevertheless, a community formed, a direction of travel was established, and a handful of great minds began to work together. In time, an eclectic collection of different types of problems would be swept together under the banner of AI: recognizing human speech, analyzing images and objects, translating and interpreting written text, playing games like checkers and chess, and problem solving.[13]

In the beginning, most AI researchers believed that building a machine to perform a given task meant observing how human beings performed the same task and copying them. At the time, this approach must have seemed entirely sensible. Human beings were by far the most capable contraptions in existence, so why not try to build these new machines in their image?

This mimicry took various forms. Some researchers tried to replicate the actual physical structure of the human brain, attempting to create networks of artificial neurons. (Marvin Minsky, one of the authors of that Rockefeller pitch asking for funding, wrote his PhD thesis on how to build these.)[14] Others tried a more psychological approach, looking to replicate the thinking and reasoning processes in which the human brain appeared to be engaged. (That was what Dartmouth attendees Herbert Simon and Allen Newell tried to do with their "General Problem Solver," an early system celebrated as "a program that simulates human thought.")[15] Yet a third approach was to draw out the rules that human beings seemed to follow, and then write instructions for machines based on those. Researchers established a dedicated subfield to do this, their handicrafts known as "expert systems"—"expert" because they relied upon the rules that a human expert gave them to use.

In all of these efforts, human beings provided the template for machine behavior in one way or another. Designing a machine that could play world-class chess, for instance, meant sitting down with grand masters and getting them to explain how they went about playing the game. Writing a program to translate one language into another meant observing how a multilingual person makes sense of a paragraph of text. Identifying objects meant representing and processing an image in the same way as human vision.[16]

This methodology was reflected in the language of the AI pioneers. Alan Turing claimed that "machines can be constructed which will simulate the behaviour of the human mind very closely."[17] Nils Nilsson, an attendee at the Dartmouth gathering, noted that most academics there "were interested in mimicking the higher levels of human thought. Their work benefitted from a certain amount of introspection about how humans solve problems."[18] And John Haugeland, a philosopher, wrote that the field of AI was seeking "the genuine article: *machines with minds*, in the full and literal sense."[19]

Behind some of the claims made by Haugeland and others was a deeper theoretical conviction: human beings, they believed, were themselves actually just a complex type of computer. This was the "computational theory of the mind." From a practical point of view, it may have been an appealing idea for AI researchers. If human beings were only complicated computers, the difficulty of building an artificial intelligence was not insurmountable: the researchers merely had to make their own, simple computers more sophisticated.[20] As the computer scientist Douglas Hofstadter puts it in his celebrated *Gödel, Escher, Bach*, it was an "article of faith" for many researchers that "all intelligences are just variations on a single theme; to create true intelligence, AI workers will have to keep pushing . . . closer and closer to brain mechanisms, if they wish their machines to attain the capabilities which we have."[21]

Of course, not everyone was interested in copying human beings. But most AI specialists of that vintage were—and even those who initially were not eventually got drawn toward it. Take Herbert Simon and Allen Newell, for example. Before they created the General Problem Solver, their system based on human reasoning, they had in fact built an entirely different system, called the Logic Theorist. This system stood out from all the others: unlike any other machine from that original Dartmouth gathering, it actually worked. And yet, despite this success, Simon and Newell abandoned it. Why? In part because it did not perform like a human being.[22]

Ultimately, however, this approach of building machines in the image of human beings did not succeed. Despite the initial burst of optimism and enthusiasm, no serious progress was made in AI. When it came to the grand challenges—building a machine with a mind, one that was conscious, or that could think and reason like a human being—the defeat was emphatic. Nothing got close. And the same was true for the more pedestrian ambitions of getting machines to perform specific tasks. Despite all the efforts, machines could not beat a top player at chess. They could not translate more than a handful of sentences or identify anything but the simplest objects. And the story was the same for a great many other tasks, too.

As progress faltered, researchers found themselves at a dead end. The late 1980s became known as the "AI winter": funding dried up, research slowed, and interest in the field fell away. The first wave of AI, that had raised so many hopes, ended in failure.

THE SECOND WAVE OF AI

Things started looking up again for AI in 1997. That's when a system called Deep Blue, owned by IBM, beat Garry Kasparov, then the world chess champion. It was a remarkable achievement, but even more remarkable was how the system did it. Deep Blue did not try to copy Garry Kasparov's creativity, his intuition, or his genius. It did not replicate his thinking process or mimic his reasoning. Instead, it used vast amounts of processing power and data storage to crunch up to 330 million moves in a second. Kasparov, one of the best human chess players of all time, could hold perhaps up to a hundred possible moves in his head at any one moment.[23]

The Deep Blue result was a practical victory, but it was an ideological triumph as well. We can think of most AI researchers until then as *purists*, who closely observed human beings acting intelligently and tried to build machines like them. But that was not how Deep Blue was designed. Its creators did not set out to copy the anatomy of human chess players, the reasoning they engaged in, or the particular strategies they followed. Rather, they were *pragmatists*, taking a task that required intelligence when performed by a human being and building a machine to perform it in a fundamentally different way. That's what brought AI out of its winter—what I call the pragmatist revolution.

In the decades since Deep Blue's victory, a generation of machines has been built in this pragmatist spirit: crafted to function very differently from human beings, judged not by *how* they perform a task but *how well* they perform it. Advances in machine translation, for instance, have come not from developing a machine that mimics a talented translator, but from having computers scan millions of human-translated pieces of text to figure out interlingual correspondences and patterns on their own. Likewise, machines have learned to classify images, not by mimicking human vision but by reviewing millions of previously labeled pictures and hunting for similarities between those and the particular photo in question. The ImageNet project hosts an annual contest where leading computer scientists compete to build systems that can identify objects in an image more accurately than their peers—and in 2015, the winning system outperformed human beings for the first time, correctly identifying images 96 percent of the time. In 2017, the winner reached 98 percent accuracy.

Like Deep Blue, many of these new machines rely on recent advances in processing power and data storage. Remember that between the 1956 Dartmouth gathering and the end of the century, there was a roughly ten billion–fold increase in the power of a typical computer. As for data, Eric Schmidt, the former chairman of Google, estimates that we now create as much information every two days as was created from the dawn of civilization to 2003.[24]

In the first wave of AI, before such processing power and massive data storage capabilities were available, people had to do much of the

Figure 3.1: Error Rate of the Winning System in the ImageNet Contest[25]

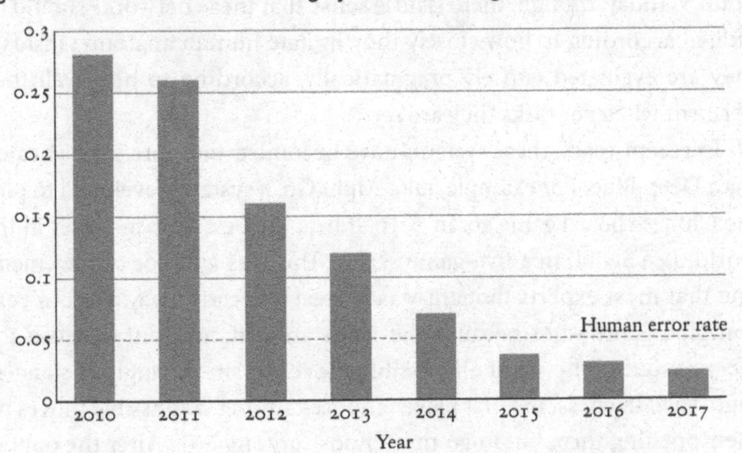

difficult computational work themselves. Researchers hoped, through their own ingenuity, insight, and introspection, to uncover the way that human beings thought and reasoned, or to manually discover the hidden rules that shaped their behavior, and to capture all of this in a set of explicit instructions for machines to follow. But in the second wave of AI, machines no longer relied on this top-down application of human intelligence. Instead, they began to use vast amounts of processing power and increasingly sophisticated algorithms to search through huge bodies of data, mining human experience and example to figure out what to do themselves, from the bottom up.[26]

The word *algorithm*, derived from the name of a ninth-century mathematician called Abdallah Muhammad ibn Mūsā Al-Khwārizmī, simply means a set of step-by-step instructions. Machine-learning algorithms, which drive much of the progress in AI today, are specifically aimed at letting systems learn from their experience instead of being guided by explicit rules. Many of them draw on ideas from the early days of AI—ideas developed long before there was enough processing power and data available to turn them from intriguing theoretical possibilities into something more practical. Indeed, some of today's greatest pragmatist triumphs have grown out of earlier purist attempts to copy human beings. For instance, many of the most capable machines today rely on what are known as "artificial neural networks," which were first built decades ago in an attempt to simulate the workings of the human brain.[27] Today, though, there is little sense that these networks should be judged according to how closely they imitate human anatomy; instead, they are evaluated entirely pragmatically, according to how well they perform whatever tasks they are set.[28]

In recent years, these systems have become even more sophisticated than Deep Blue. For example, take AlphaGo, a system developed to play the Chinese board game go. In 2016, it beat the best human player in the world, Lee Sedol, in a five-game series. This was a major achievement, one that most experts thought was at least a decade away. Go is a very complex game—not because the rules are difficult, but because the "search space," the set of all possible moves to sift through, is staggeringly vast. In chess, the first player chooses among 20 possible moves on their opening turn, but in go they choose among 361. After the opposing player has responded, there are 400 possible positions in chess, but around 129,960 in go. After two moves apiece, that number rises to 71,852 possible positions in chess, but about 17 billion in go. And after only three moves each, there are about 9.3 million possibilities in chess, but about 2.1×10^{15} in go—a two followed by fifteen zeros. That's 230 million times as many possibilities as in chess at that same early point in the game.[29]

In chess, Deep Blue's victory came in part from its ability to use brute-force processing power to calculate further ahead in a game than Kasparov could. But because of go's complexity, that strategy would not work for AlphaGo. Instead, it took a different approach. First it reviewed 30 million moves from games played by the best human experts. Then it learned from playing repeatedly against itself, crunching through thousands of games and drawing insights from those, too. In this way, AlphaGo was able to win while evaluating far fewer positions than Deep Blue had done in its matches.

In 2017, a yet more sophisticated version of the program was unveiled, called AlphaGo Zero. What made this system so remarkable is that it had wrung itself dry of any residual role for human intelligence altogether. Buried within Deep Blue's code were still a few clever strategies that chess champions had worked out for it to follow in advance.[30] And in studying that vast collection of past games by great human players, AlphaGo was in a sense relying on them for much of its difficult computational work. But AlphaGo Zero required none of this. It did not need to know anything about the play of human experts; it did not need to try to mimic human intelligence at all. All it needed was the rules of the game. Given nothing more than those, it played itself for three days to generate its own data—and it returned to thrash its older cousin, AlphaGo.[31]

Other systems are using similar techniques to engage in pursuits that more closely resemble the messiness of real life. Chess and go, for instance, are games of "perfect information": both players see the entire board and all the pieces. But as the legendary mathematician John von Neumann put it, "real life is not like that. Real life consists of bluffing, of little tactics of deception, of asking yourself what is the other man going to think I mean to do." That is why poker has fascinated researchers—

and proven so hard to automate. Yet DeepStack, developed by a team in Canada and the Czech Republic in 2017, managed to defeat professional poker players in a series of more than forty-four thousand heads-up games (that is, games involving two players). Like AlphaGo Zero, it did not derive its tactics from reviewing past games played by human experts. Nor did it rely on "domain knowledge," clever predetermined poker strategies set down by human beings for it to follow. Instead, it learned to win just by exploring several million randomly generated games.[32] In 2019, Facebook and Carnegie Mellon University went a step further: they announced the creation of Pluribus, a system that could defeat the best professional poker players in *multiplayer* competitions as well. This system, too, learned poker purely "from scratch," without any human input, just by playing hand after hand against copies of itself over the course of a few days.[33]

A SHIFT IN PRIORITY

It would be a mistake to think that researchers simply "discovered" the pragmatist path in a eureka moment early in the 1990s. The choice between trying to have machines solve problems by any means possible and having them specifically copy the human approach, between what I call "pragmatism" and "purism," is not particularly new. Back in 1961, in the heyday of purism, Allen Newell and Herbert Simon wrote that researchers were drawing "a careful line . . . between the attempt to *accomplish* with machines the same tasks that humans perform, and the attempt to *simulate* the processes humans actually use to accomplish these tasks."[34] In 1979, Hubert Dreyfus, a philosopher of AI, similarly drew a distinction between "AI engineers"—a practical bunch "with no interest in making generally intelligent machines"—and "AI theorists," engaged in what he saw as a virtuous scholarly calling.[35] Early researchers were aware that they could focus either on building machines that performed tasks that historically had required intelligent human beings, or on trying to understand human intelligence itself.[36] For them, it was the latter that really captured their attention.

Partly, as we have seen, this was because copying human beings initially seemed like the path of least resistance for building capable machines. Humans had remarkable capabilities; why start from scratch when designing a machine when you could simply mimic the human instead? But there was also a second reason for their inclination toward purism. For many researchers, the project of understanding human intelligence for its own sake was simply a lot more interesting than merely building capable machines. Their writings are full of excited references to classical thinkers and their reflections on the human mind—people like Gottfried Wilhelm Leibniz and Thomas Hobbes, René Descartes and David Hume. These were the figures in whose footsteps AI researchers hoped they were following. They were thrilled by questions not about machines, but about human beings—what is a "mind" and how does "consciousness" work, what does it really mean to "think" or to "understand."

Artificial intelligence, for many pioneers of the field, was only ever a mechanical means to that human end. As the influential philosopher John Searle put it, the only purpose of AI was as a "power tool in the study of the mind." Hilary Putnam, another prominent philosopher, thought the field of AI should be judged solely by whether it "taught us anything of importance . . . about how we think."[37] This is why many AI researchers at the time thought of themselves as *cognitive* scientists, not *computer* scientists at all.[38] They imagined they were working in a subfield of a far bigger project: understanding the human brain.[39]

Today, though, a shift in priority is taking place. As technological progress has accelerated, it has become clear that human intelligence is no longer the only route to machine capability. Many researchers now are interested less in trying to understand human intelligence than in building well-functioning machines. What's more, the researchers who remain interested in the pursuit of an understanding of human intelligence for its own sake are also finding themselves dragged toward the more pragmatic ends. The capabilities of the pragmatists' machines have aroused the interest of large technology companies—Google, Amazon, Apple, Microsoft, and others. With access to massive amounts of data and processing power, and with throngs of talented researchers on their payroll, these companies have taken the lead in AI development, shaping the field and changing its priorities. To many of them, the pursuit of an understanding of human intelligence for its own sake must look like an increasingly esoteric activity for daydreaming scholars. In order to stay relevant, many researchers—even those inclined to the purist side—have had to align themselves more closely with these companies and their commercial ambitions.

Take DeepMind, for example, the British AI company that developed AlphaGo. It was bought by Google in 2014 for $600 million, and is now staffed by the leading minds in the field, poached from top academic departments by pay packages that would make their former colleagues blush—an average of $345,000 per employee.[40] The company's mission statement says that it is trying "to solve intelligence," which at first glance suggests they might be interested in figuring out the puzzle of the human brain. But look more closely at their actual achievements, and you can see that in practice their focus is quite different. Their machines, such as AlphaGo, are only "intelligent" in the sense that they are very capable, in some cases remarkably so—but they do not think or reason like human beings.

In the same way, think of AI assistants such as Amazon's Alexa, Apple's Siri, and Microsoft's Cortana. We can pose simple questions to these machines, as we would to a human being, and they will answer in a relatively compelling human voice. (In 2018, Google showed off a demo recording of Duplex, their AI assistant, calling a hair salon to book an appointment—the receptionist had no idea she was talking to a machine, so realistic was its intonation and humanlike "mmm-hmms.")[41] Yet however impressive systems like this may be, however closely they may resemble human beings, they are not exhibiting intelligence akin to that of people. Their inner workings do not resemble the mind. They are not conscious. They do not think, feel, or reason like people do.

Is it appropriate, then, to describe these machines as "intelligent" in any sense at all? It doesn't feel quite right. Most of the time, we end up relying on that word, or similar ones, because we do not have anything else that works. But because we mainly use those words to talk about human beings, it feels wrong to use them to talk about machines as well. Philosophers would call this a "category mistake," using a word meant for one category of things in another: just as we would not expect a carrot to talk or a mobile phone to be angry, we should not expect a machine to be "intelligent" or "smart." How else might we describe them, then? When the field of AI was just starting up, and it didn't yet have a name, one thought was to call the area of research "computational rationality." This term may not be quite as exciting or provocative as "artificial intelligence," but it is probably a better match, since this is exactly what these machines are doing: using computational power to search through a vast ocean of possible actions for the most rational one to take.

BOTTOM-UP, NOT TOP-DOWN

In many ways, the current pragmatist revolution in AI resembles another intellectual revolution that has taken place over the past century and a half, one that has shaped how we think about the intellectual capabilities of a different machine: the human being.

Consider that until now, human beings, armed with their intelligence, have been the most capable machines in existence. For a long time, the dominant explanation for the origins of those capabilities was religious: they came from God, from something even more intelligent than us crafting us in its image. How, after all, could such a complicated machine as a human being exist if something vastly more intelligent than us had not designed it to be that way? William Paley, an eighteenth-century theologian, asked us to imagine ourselves walking through a field. If we stumble upon a rock lying in the grass, he said, we might think it has been there forever, but if we find a wristwatch, we would not think the same thing. A device as complex as a watch could not have been there since time began, he argued. At some point, a talented watchmaker must have designed and manufactured it. All the complex things we find in nature, thought Paley, are like that watch: the only way to explain them is with a creator, a watchmaker somewhere out of sight, composing plans and putting them into action.

The similarities between these religious scholars and the AI purists are striking. Both were interested in the origins of machine capabilities—the former, human machines, the latter, man-made ones. Both believed that these had to be deliberately created by an intelligence that closely resembled their own, a so-called intelligent designer. For the religious scholars, that designer was God. For the AI purists, that designer was, well, themselves. And both were convinced that the creations ought to be like their creator. Just as the God of the Old Testament created man in His image, the AI researchers tried to build their machines in their own image, too.[42]

In sum, both the theologians and the AI scientists believed that remarkable capabilities could only ever emerge from something that resembled human intelligence. In the words of the philosopher Daniel Dennett, both thought that *competence* could only emerge from *comprehension*, that only an intelligent process could create exceptionally capable machines.[43]

Today, though, we know that the religious scholars were wrong. Humans and human capabilities were not created through the top-down efforts of something more intelligent than us, molding us to look like it. In 1859, Charles Darwin showed that the reverse was true: the creative force was a bottom-up process of unconscious design. Darwin called this "evolution by natural selection," the simplest account of it only requiring you to accept three things: first, that there are slight variations between living beings; second, that some of these variations might be favorable for their survival; and third, that these variations are passed on to others. There was no need for an intelligent designer, directly shaping events; these three facts alone could explain all appearances of design in the natural world. The variations might be tiny, the advantages ever so slight, but these changes, negligible at any instant, would—if you left the world to run for long enough—accumulate over billions of years to create dazzling complexity. As Darwin put it, even the most "complex organs and instincts" were "perfected, not by means superior to, though analogous with, human reason, but by the accumulation of innumerable slight variations, each good for the individual possessor."[44]

The ideas of natural selection and intelligent design could not be more different. "A true watchmaker has foresight," wrote Richard Dawkins, one of the great scholars of Darwin. "He designs cogs and springs, and plans their interconnections with a future purpose in mind's eye. Natural selection, the blind unconscious, automatic process which Darwin discovered . . . has no purpose in mind. It has no mind and no mind's eye. It does not plan for the future. It has no vision, no foresight, no sight at all. If it can be said to play the role of watchmaker in nature, it is the *blind* watchmaker."[45]

The watchmaker of Paley's story had perfect vision and foresight, but Darwin's process of natural selection has none of that. It is oblivious to what it does, mindlessly stumbling upon complexity across eons rather than consciously creating it in an instant.

The pragmatist revolution in AI requires us to make a similar reversal in how we think about where the abilities of man-made machines come from. Today, the most capable systems are not those that are designed in a top-down way by intelligent human beings. In fact, just as Darwin found a century before, remarkable capabilities can emerge gradually from blind, unthinking, bottom-up processes that do not resemble human intelligence at all.[46]

4

Underestimating Machines

In 1966, during the first wave of AI, a researcher named Joseph Weizenbaum announced that he had built ELIZA, the world's first chatbot: a system that would "make certain kinds of natural language conversation between man and computer possible."[1] At first, ELIZA was set up to act as a psychotherapist. The idea was that a "patient" would speak to it, the system would respond with a comment on what the patient had said, and conversation would then flow back and forth. Weizenbaum did not really intend for it to be taken very seriously, though. In part, he seemed to be poking fun at the predictable way that psychotherapists tend to parrot back a version of the patient's own statements with an air of measured profundity. It was, he wrote, meant to be a "parody."

Once ELIZA was up and running, though, events took an unexpected turn. The system made a far more powerful impression on its users than Weizenbaum had anticipated. Some practicing psychiatrists who reviewed the system thought it might "be ready for clinical use" after a bit of further work. When Weizenbaum invited his own secretary, who knew full well how ELIZA was built, to try it out, she turned to him after a brief initial exchange and asked him to leave the room: she wanted to spend some time alone with the machine. Weizenbaum was shocked. A few years later he wrote that the experience had "infected" him with "fundamental questions [that] I shall probably never be rid of . . . about man's place in the universe."[2]

Weizenbaum did not expect very much from ELIZA because he knew that while the system might outwardly appear to be intelligent, it did not in fact think or feel like a human therapist at all. "I had thought it essential," he wrote, "as a prerequisite to the very possibility that one person might help another learn to cope with his emotional problems, that the helper himself participate in the other's experience of those problems."[3] But by dismissing his system on those grounds, Weizenbaum underestimated the capabilities of what he had created.

When the pragmatist revolution began to unfold a few decades later, and researchers started to systematically build machines that worked in very different ways from human beings, Weizenbaum's mistake would be made far more regularly and with more serious consequences. AI researchers, economists, and many others would be caught out, time and time again, by the capabilities of new machines that were no longer built to copy some supposedly indispensable feature of human intelligence.

A SENSE OF DISAPPOINTMENT

For an influential group of critics in the AI community, the pragmatist revolution is more a source of disappointment than a cause for celebration. Take their response to the chess triumph of IBM's Deep Blue over Garry Kasparov. Douglas Hofstadter, the computer scientist and writer, called its first victory "a watershed event" but dismissed it as something that "doesn't have to do with computers becoming intelligent."[4] He had "little intellectual interest" in IBM's machine because "the way brute-force chess programs work doesn't bear the slightest resemblance to genuine human thinking."[5] John Searle, the philosopher, dismissed Deep Blue as "giving up on A.I."[6] Kasparov himself effectively agreed, writing off the machine as a "$10 million alarm clock.[7]

Or take Watson, another IBM computer system. Its claim to fame is that in 2011, it appeared on the US quiz show *Jeopardy!* and beat the two top human champions of the show. In the aftermath, Hofstadter again agreed that the system's performance was "impressive" but also said that it was "absolutely vacuous."[8] Searle, in a sharp *Wall Street Journal* editorial, wryly noted, "Watson didn't know it won on *Jeopardy!*"[9] Nor did the machine want to call up its parents to say how well it had done, or go to the pub to celebrate with its friends.

Hofstadter, Kasparov, Searle, and those who make similar observations are all correct, as we saw in the last chapter. Despite endless talk by businesses and the media about "artificial intelligence" or "machine intelligence," actual intelligence is not really what is emerging. Purists like Searle and Hofstadter wanted to use AI research to solve the puzzle of human intelligence, to unravel the mystery of consciousness and learn about the mind—yet today's most capable machines cast little light on human functioning. And so, understandably, they feel let down.

By itself, this disappointment is not unreasonable. The trouble comes when it tips into disparagement, as it often tends to do. Some of these critics seem to think that because the latest machines do not think like intelligent human beings, they are unimpressive or superficial—and this has led them to systematically underestimate the machines' capabilities.

This way of thinking explains why these critics have often found themselves trapped in a recurring pattern of intellectual evasion. When some task gets automated that they had thought could only be performed by human beings, they respond by saying that this task was not a proper reflection of human intelligence after all—and then gesture toward some entirely different task, one that cannot yet be automated, and argue that *that* is actually what human intelligence is all about. Religious leaders are sometimes criticized for defining "God" as whatever science cannot currently explain, a "God of the gaps": God is the power that made night and day—until astronomy could explain them; God is the power that made all living creatures—until evolution was understood to be responsible. Here, we have a similarly fluid style of definition at work, an "intelligence of the gaps," where intelligence is defined as whatever machines currently cannot do. Being aware of this trap does not necessarily protect you from plunging into it, either. Hofstadter, for instance, was well aware of the phenomenon, wittily identifying it himself as a "theorem" of the field: "AI is whatever hasn't been done yet."[10] Yet, having described the fallacy himself, he still fell for it.

Back in 1979, asking himself whether there would ever be a chess program that "can beat anyone," Hofstadter had bluntly replied: "No. There may be programs which can beat anyone at chess, but they will not be exclusively chess players. They will be programs of *general* intelligence, and they will be just as temperamental as people. 'Do you want to play chess?' 'No, I'm bored with chess. Let's talk about poetry.' That may be the kind of dialogue you could have with a program that could beat everyone."[11] In other words, Hofstadter thought that a successful chess-playing system would have to have human intelligence. Why? Because he was a purist. He believed that "chess-playing draws intrinsically on central facets of the human condition," such as the "ability to sort the wheat from the chaff in an intuitive flash, the ability to make subtle analogies, the ability to recall memories associatively."[12]

As we have seen, though, Deep Blue showed this was wrong: none of that human magic, no intuitive sorting of wheat from chaff, was necessary to play sensational chess. But rather than admit a mistake, Hofstadter went for the "intelligence of the gaps" evasion. These machines are "just overtaking humans in certain intellectual activities that we thought required intelligence," he wrote after Deep Blue won its first game against Kasparov. "My God, I used to think chess required thought. Now, I realize it doesn't. It doesn't mean Kasparov isn't a deep thinker, just that you can bypass deep thinking in playing chess, the way you can fly without flapping your wings."[13] Hofstadter had changed his mind, denying that the capabilities required to play chess were an "essential ingredient" of human intelligence after all.[14]

Or take Kasparov himself, the human protagonist in this story of man versus chess-playing machine. In *Deep Thinking*, his account of those matches with Deep Blue, he also identifies the trap: "as soon as we figure out a way to get a computer to do something intelligent, like play world championship chess, we decide it's not truly intelligent."[15] Yet this is exactly what he has done. Seven years before he faced Deep Blue, Kasparov had boldly claimed that a machine could never beat someone like him, since it could never be human like him: "If a computer can beat the world champion, [then] the computer can read the best books in the world, can write the best plays, and can know everything about history and literature and people. That's impossible."[16] To Kasparov back then, winning at chess was inseparable from everything else that makes us who we are. Yet after the match, he would claim that "Deep Blue . . . was only intelligent in the way your programmable alarm clock is intelligent."[17] Like Hofstadter, he changed his mind, deciding that winning at chess was not after all a sign of human intelligence.

This habit of shifting the goalposts is unhelpful because it tends to lead critics to underestimate the capabilities of machines yet to come. And even on its own terms, you might wonder whether this dismissive tone is misplaced. What is so uniquely remarkable about human intelligence? Why do we elevate human thinking, astounding though it may be, above any other approach that can create remarkable machine capabilities? Of course, the power and mystery of the human mind should leave us in awe. We may not fully understand the inside of our heads for quite some time. But is there not also another sense of amazement, no less unsettling and thrilling, in the designs of machines that, even if they do not resemble or replicate human beings, can outperform us? Kasparov may dismiss Deep Blue as just an expensive alarm clock, but this alarm clock gave him a pounding at the chessboard. Should its inner workings not amaze us like those of the brain, even if they do not share our wondrous anatomy and physiology?

After all, that is how Darwin felt when he realized that the capabilities of the *human* machine did not emerge from something that resembles human intelligence.[18] He was not a bitter man, trying to drain any last sense of magic or mystery from the world with his creator-less theory of natural selection. Quite the contrary. Consider the final words of *On the Origin of Species*: "There is grandeur in this view of life, with its several powers, having been originally breathed into a few forms or into one; and that, whilst this planet has gone cycling on according to the fixed law of gravity, from so simple a beginning endless forms most beautiful and most wonderful have been, and are being, evolved."[19]

This is not the writing of a metaphysical grinch. Darwin's view of life without a creator has a "grandeur" to it, and is articulated with an almost religious sense of awe. One day we may feel that way about our unhuman machines as well.

ARTIFICIAL GENERAL INTELLIGENCE

The ancient Greek poet Archilochus once wrote: "The fox knows many things, but the hedgehog knows one big thing." Isaiah Berlin, who found this mysterious line in the surviving scraps of Archilochus's poetry, famously used it as a metaphor to distinguish between two types of human being: people who know a little about a lot (the foxes) and people who know a lot about a little (the hedgehogs).[20] In our setting, we can repurpose that metaphor to think about human beings and machines. At the moment, machines are prototypical hedgehogs, each of them designed to be very strong at some extremely specific, narrowly defined task—think of Deep Blue and chess, or AlphaGo and go—but hopeless at performing a range of different tasks. Human beings, on the other hand, are proud foxes, who might now find themselves thrashed by machines at certain undertakings, but can still outperform them at a wide spread of others.

For many AI researchers, the intellectual holy grail is to build machines that are foxes rather than hedgehogs. In their terminology, they want to build an "artificial general intelligence" (AGI), with wide-ranging capabilities, rather than an "artificial narrow intelligence" (ANI), which can only handle very particular assignments.[21] That is what interests futurists like Ray Kurzweil and Nick Bostrom. But there has been little success in that effort, and critics often put forward the elusiveness of AGI as a further reason for being skeptical about the capabilities of machines. There is a sense among purists that only AGI is "real" AI, and that without this generality of capability these machines will never be "true rivals" to human beings in the work that they do.[22]

AGI, it is said, will represent a turning point in human history—perhaps *the* turning point. The idea is that once machines have "general" capabilities, and are able to perform a wide range of tasks better than human beings can, then it is only a matter of time before the task of designing yet more capable machines falls within their reach. At this point, it is thought, an "intelligence explosion" will take place: machines endlessly improving upon those that came before, their capabilities soaring in an ever-accelerating blast of recursive self-improvement. This process, it is said, will lead to machines with "superintelligence";

some call it the "singularity." These machines would be the "last invention that man need ever make," wrote Irving John Good, the Oxford mathematician who introduced the possibility of such an intelligence explosion: anything a human being could invent, they could improve upon.[23]

The prospect of such vastly capable AGIs has worried people like Stephen Hawking ("could spell the end of the human race"), Elon Musk ("vastly more risky than North Korea"), and Bill Gates ("don't understand why some people are not concerned")—though their worries are not always the same.[24] One fear is that human beings, limited in what they can do by the comparatively snaillike pace of evolution, would struggle to keep up with the machines. Another is that these machines might, perhaps unwittingly, pursue goals at odds with those of human beings, destroying us in the process. One thought experiment, for example, imagines an AGI that is tasked with manufacturing paper clips as efficiently as it can; the story ends with it turning "first all of earth and then increasing portions of space into paperclip manufacturing facilities," trampling over humans in the ruthlessly successful pursuit of its set goal.[25]

Experts are divided on how long it might take before we actually get there. Some say AGIs are a few decades away, others say more like centuries; a recent survey converged, with improbable precision, on 2047.[26] Today, we do see some small steps in the direction of "general" capabilities, although these are just very early and primitive examples of it at work. As part of its portfolio of innovations, for instance, DeepMind has developed a machine that is able to compete with human experts at forty-nine different Atari video games. The only data this machine receives is the pattern of pixels on the computer screen and the number of points it has won in the game; yet even so, it has been able to learn how to play each distinct game, often to a level that rivals the finest human players.[27] This is the sort of general capability that AGI enthusiasts are chasing after.

Discussions like this, about "intelligence explosions" and "superintelligence," might be thrilling. But in thinking about the future of work, the importance of AGI is greatly exaggerated compared to that of ANI. For AI researchers, the absence of AGI is a pressing bottleneck; but in economics, it is a far weaker constraint on automation than commonly imagined. If a job is made up of ten tasks, for instance, there are two ways that progress in AI could make it disappear. One is that an AGI is created that can perform all ten tasks by itself; the other is that ten distinct ANIs are invented, each able to perform just one of the tasks involved. Our fascination with AGI, with building machines that have general capabilities like human beings, risks distracting us from quite how powerful machines can be without it. It is not necessary to build a single machine in the image of a human being that can displace workers in an instant. Instead the gradual accumulation of a range of unhuman machines, with narrow but impressive abilities, is enough to erode the individual tasks that people carry out. In short, when thinking about the future of work, we should be wary not of one omnipotent fox, but an army of industrious hedgehogs.

REVISITING THE ECONOMISTS

The pragmatist revolution in AI has also had serious consequences for economists. In the last few years, it has caused the ALM hypothesis to break down.

When David Autor and his colleagues first developed the ALM hypothesis back in 2003, it was accompanied by a list of "non-routine" tasks. The authors were confident that these tasks could not be readily automated—yet today, the majority of them can be. One was "driving a truck," but Sebastian Thrun developed the first driverless vehicle the following year. Another was "legal writing," yet document automation systems are now commonplace in most major practices. "Medical diagnosis" was also thought to be safe, but today machines can detect eye problems and identify cancers, among much else.[28]

A decade later, Autor and another colleague identified "order-taking" as "non-routine." But that same year, Chili's and Applebee's restaurants in the United States announced they were installing one-hundred-thousand tablets to allow customers to order and pay without a human waiter; McDonald's and other chains have followed suit. Even some highly esoteric "non-routine" tasks have fallen. Just a few years ago it was claimed that "identifying a bird based on a fleeting glimpse" could not readily be automated, yet there is a system now that does precisely this—a program called Merlin, designed by computer scientists at the Cornell Laboratory of Ornithology in the US.

Of course, it is reasonable to question whether these tasks have been entirely automated. At the moment, there are still ailments that diagnostic systems are powerless to interpret, and birds that Merlin cannot identify. Today's "autopilot" car systems still require human attention. But it is important to notice the direction of travel: many "non-routine" tasks are now within the grasp of machines, something that was unthinkable until very recently.[29]

What went wrong with the predictions? The problem is that the ALM hypothesis neglected the pragmatist revolution. Economists had thought that to accomplish a task, a computer had to follow explicit rules articulated by a human being—that machine capabilities had to begin with the top-down application of human intelligence. That may have been true in the first wave of AI. But as we have seen, it is no longer the case. Machines can now learn how to perform tasks themselves, deriving their own rules from the bottom up. It does not matter if human beings cannot readily explain how they drive a car or recognize a table; machines no longer need those human explanations. And that means they are able to take on many "non-routine" tasks that were once considered to be out of their reach.

The immediate implication is clear: economists have to update the stories they tell about technology and work. The set of tasks that remain for human beings to do has shrunk beyond the boundaries they once imagined were in place. And in the latest work done by leading economists, such revisions have begun. There is a growing realization that the traditional assumptions made about machine capabilities no longer hold, that something has gone awry with the "routine" and "non-routine" distinction. Yet the response has still often been to keep that original distinction, merely tweaking and updating it, instead of abandoning the ALM hypothesis altogether.

At the moment, many economists believe their mistake was a failure to see that new technologies would turn many "non-routine" tasks into "routine" ones. Remember that "non-routine" tasks are defined as those requiring "tacit" knowledge, the sort of knowledge that human beings struggle to articulate. What is happening, the economists argue, is that new technologies are simply uncovering some of this tacit knowledge that human beings rely upon. The "routine" versus "non-routine" distinction is still useful, they maintain; new technologies are simply shifting the boundary between the two, a boundary that the economists once (mistakenly, they admit) thought was fixed. In this way, they are trying to rescue the ALM hypothesis. David Autor, for instance, argues that today's computer scientists are trying to automate "non-routine" tasks by "inferring the rules that we tacitly apply but do not explicitly understand"; Dana Remus and Frank Levy say that new technology "makes the tacit protocol explicit."[30]

To see this way of thinking in action, consider a system that was developed by a team of researchers at Stanford University in 2017 to detect skin cancer. If you give it a photo of a freckle, it can tell whether it is cancerous as accurately as twenty-one leading dermatologists. How does it work? It draws on a database of 129,450 past cases, hunting for similarities between those cases and an image of the particular lesion in question. The updated ALM hypothesis suggests that this freckle machine works because it is able to identify and extract from those past cases the ineffable rules that dermatologists follow but cannot themselves articulate. The machine is making their tacit rules explicit, turning a "non-routine" task into a "routine" one.

But this explanation of how the Stanford freckle-analyzing machine works is incorrect. The idea that such machines are uncovering hitherto hidden human rules, plunging deeper into people's tacit understanding of the world, still supposes that it is *human intelligence* that underpins machine capability. But that misunderstands how second-wave AI systems operate. Of course, some machines may indeed stumble upon unarticulated human rules, thereby turning "non-routine" tasks into "routine" tasks. But far more significant is that many machines are also now deriving entirely *new* rules, unrelated to those that human beings follow. This is not a semantic quibble, but a serious shift. Machines are no longer riding on the coattails of human intelligence.

Take the Stanford freckle machine again. When it searches through those 129,450 past clinical cases, it is not trying to uncover the "tacit" rules that a dermatologist follows. It is using massive amounts of processing power to identify crucial patterns in a database containing more possible cases than a doctor could hope to review in her lifetime. Of course, it may be that some of the rules it uncovers from that exercise resemble those that human beings follow. But that is not necessarily so: the machine may also discover entirely different rules, too, ones that human beings do not follow at all.

Another good example of this is AlphaGo, the go-playing machine that beat the world champion Lee Sedol. Almost as remarkable as its overall victory was a particular move that AlphaGo made—the thirty-seventh move in the second game—and the reaction of those watching. The commentators were shocked. They had never seen a move like it. Lee Sedol himself appeared deeply unsettled. Thousands of years of human play had forged a rule of thumb known even to beginners: early in the game, avoid placing stones on the fifth line from the edge. And yet, this is exactly what AlphaGo did in that move.[31] The system had not discovered an existing but hitherto unarticulated human rule. In fact, AlphaGo itself calculated from the data at its disposal that the probability of a human expert playing that move was just one in ten thousand.[32] As one expert observer noted, it was "not a human move" at all.[33] It was new and astonishing. One go champion called it "beautiful." Another said it made him feel "physically unwell."[34] This system quite literally rewrote the rules that human beings follow.

It might be tempting to disregard this distinction, to say that all that really matters is for economists to recognize that "non-routine" tasks can now be automated. Yet the reason *why* these economists were wrong matters as well. The very fact that these systems may not follow the same rules as human beings creates opportunities for them, as with AlphaGo surprising Lee Sedol—and it creates problems as well. Consider, for instance, that one of the merits of systems developed during the first wave of AI was that they were very "transparent." Since they tended to follow explicit rules set down by human beings, it was easy for someone to understand why a system reached any given decision, whether a move in a game or a medical diagnosis. In the second wave of AI, this is no longer the case; these systems are now far more "opaque." Why AlphaGo chose to play that unprecedented thirty-seventh move, for instance, was initially not clear at all, and the system's developers had to carefully review the complex calculations it had made before they could make sense of its decision.

This new opacity has spurred a dedicated research effort to help AI systems "explain themselves."[35] It has also already provoked a tentative public policy response. In the EU, for example, Article 15 of the new General Data Protection Regulation has made "meaningful information about the logic [of] automated decision-making" a legal right.[36] At the moment, the sense among European policymakers is that this information is missing.

THE ARTIFICIAL INTELLIGENCE FALLACY

In the various cases described above, computer scientists and economists were committing what my father and I have come to call the "AI fallacy": the mistaken belief that the only way to develop machines that perform a task at the level of human beings is to copy the way that human beings perform that task.[37] This fallacy remains widespread to this day, shaping how many people still think about technology and work.

Doctors, for instance, tend to resist the claim that a machine will ever be able to diagnose illnesses as well as they do. A machine, they say, will never be able to exercise "judgment." Judgment requires instinct and intuition, an ability to look a patient in the eye, to exercise a personal touch refined through experience; none of that could ever be written down in a set of instructions for a machine. The Royal College of General Practitioners (RCGP), the professional body for doctors in the UK, proclaims that "no app or algorithm will be able to do what a GP does . . . research has shown GPs have a 'gut feeling' when they just know something is wrong with a patient."[38] That may indeed be the case. Yet this does not mean that machines cannot possibly perform doctors' tasks: they might be able to do so by going about them in some entirely different way. The freckle-analyzing system developed at Stanford does not replicate doctors' exercise of "judgment." It does not try to replicate their "gut feeling"; indeed, one might argue that it does not "understand" anything about dermatology at all. Yet it can still tell whether a freckle is cancerous.

Architects, meanwhile, might say a machine can never design an innovative or impressive building because a computer cannot be "creative." Yet consider the Elbphilharmonie, a new concert hall in Hamburg, which contains a remarkably beautiful auditorium composed of ten thousand interlocking acoustic panels. It is the sort of space that makes one instinctively think that only a human being—and a human with a remarkably refined creative sensibility, at that—could design something so aesthetically impressive. Yet the auditorium was, in fact, designed algorithmically, using a technique known as "parametric design." The architects gave the system a set of criteria (for instance, that the space had to have certain acoustic properties, or that any panels within reach of an audience member had to have a particular texture to the touch), and it generated a set of possible designs for the architects to choose from. Similar software has been used to design lightweight bicycle frames and sturdier chairs, among much else.[39] Are these systems behaving "creatively"? No, they are using lots of processing power to blindly generate assorted possible designs, working in a very different way from a human being.

Or take one more example. In 1997, only a few months after Deep Blue beat Kasparov at chess, there was a second, largely neglected victory for AI. A roomful of listeners at the University of Oregon heard a piano piece that, they decided, was a genuine composition by Bach—but it had actually been written by EMI, a computer program invented by composer David Cope. Was this machine acting "creatively" in putting together the piece? A music theory professor at the university found the whole business "disconcerting."[40] Douglas Hofstadter, who organized the musical experiment, called EMI "the most thought-provoking project in artificial intelligence that I have ever come across," saying that it left him "baffled and troubled."[41] If a human being had written the piece, we would not hesitate to use the word *creative*. But however beautiful the composition was, it still feels wrong to use that term to describe what the computer program had done. EMI did not, as Hofstadter once wrote about composing, "have to wander around the world on its own, fighting its way through the maze of life and feeling every moment of it" before it sat down to turn those feelings into notes.[42] Once again, the machine performed its task in a very different way.

The temptation is to say that because machines cannot reason like us, they will never exercise judgment; because they cannot think like us, they will never exercise creativity; because they cannot feel like us, they will never be empathic. And all that may be right. But it fails to recognize that machines might still be able to carry out tasks that require empathy, judgment, or creativity *when done by a human being*—by doing them in some entirely other fashion.

THE FALL OF INTELLIGENCE

In Greek mythology, the ancient gods lived on top of Mount Olympus. Endowed with remarkable capabilities, they sat on the summit and looked down on the ordinary people below. If the mortals at the bottom were exceptionally valiant or distinguished, however, they, too, could become like gods: in a process the Greeks called *apotheosis*, they would ascend the mountain and take their seat on the peak. This is what happened to the Greek hero Heracles, for example. At the end of his life, he was carried up Olympus to live alongside the deities, left to dwell "unharmed and ageless for all his days."[43]

Today, many of us seem to imagine that human beings sit on top of their own mountain. We do not think we are gods, but we do consider ourselves more capable than any other creature in existence. A lot of people have assumed that, if a machine at the bottom of the mountain is to join us at the summit, it must go through apotheosis as well—not to become more like a god, but to become more like a human being. This is the purist view of AI. Once the machine gains "human intelligence," peak capability is reached and its climb is over.

But as the pragmatist revolution has shown us, there are two problems with this assumption. The first is that there are other ways to climb the Capability Mountains than to follow the particular path that human beings have taken. The purist route is just one way to make the ascent; technological progress has revealed a range of other promising paths as well. The second revelation is that there are other peaks in this mountain range alongside the one that humans proudly sit atop of. Many humans have become distracted by the view down from the summit: we spend our time looking down at the less capable machines below, or gazing at each other and marveling at our own abilities. But if we looked up, rather than down or across, we would see other mountains towering above us.

For the moment, human beings may be the most capable machines in existence—but there are a great many other possible designs that machines could take. Imagine a cosmic warehouse that stores all those different combinations and iterations: it would be unimaginably big, perhaps infinitely so. Natural selection has searched one tiny corner of this vast expanse, spent its time browsing in one (albeit very long) aisle, and settled upon the human design. However, human beings, armed with new technologies, are now exploring others. Where evolution used time, we use computational power. And it is hard to see how, in the future, we will not stumble across different designs, entirely new ways of building machines, ones that will open up peaks in capability well beyond the reach of even the most competent human beings alive today.[44]

If machines do not need to copy human intelligence to be highly capable, the vast gaps in science's current understanding of intelligence matter far less than is commonly supposed. We do not need to solve the mysteries of how the brain and mind operate to build machines that can outperform human beings. And if machines do not need to replicate human intelligence to be highly capable, there is no reason to think that what human beings are currently able to do represents a limit on what future machines might accomplish. Yet this is what is commonly supposed—that the intellectual prowess of human beings is as far as machines can ever reach.[45] Quite simply, it is implausible in the extreme that this will be the case.

PART II

THE THREAT

5

Task Encroachment

How should we expect progress in AI to affect the employment of human beings? Although machines can now do far more than in the past, this does not mean they can do everything. There are still limits to the harmful substituting force. The trouble, though, is that these boundaries are unclear and continuously changing.

Scores of recent books, articles, reviews, and reports have sought to work out the new limits of machine capabilities, using a variety of different approaches. One is to try to identify which particular human faculties are hard to automate. A popular finding, for instance, is that new technologies struggle to perform tasks that require social intelligence: activities that require face-to-face interaction or empathetic support. From 1980 to 2012, jobs that require a high level of human interaction grew by 12 percent as a share of the US workforce.[1] A 2014 Pew Research Center survey found that many experts still believed—despite all the advances of the pragmatist revolution—that there are certain "uniquely human characteristics such as empathy, creativity, judgement, or critical thinking" that will "never" be automated.[2]

A different tack, rather than looking at human faculties and asking whether they can be replicated by a machine, is to consider the tasks themselves and ask whether they have features that make them easier or harder for a machine to handle. For instance, if you come across a task where it is easy to define the goal, straightforward to tell whether that goal has been achieved, and lots of data for the machine to learn from, then that task can probably be automated.[3] Identifying photos of cats is a good example.[4] The goal is simple: just answer the question, "Is this a cat?" It is easy to tell whether a system has succeeded: "Yes, that is indeed a cat." And there are lots of photos of cats out there on the Internet, perhaps disturbingly so (about 6.5 billion of them, by one estimate).[5] By contrast, tasks with ambiguous goals or a shortage of available data may sit out of reach of machines. Economists at the Federal Reserve, the central bank of the United States, have suggested that "task complexity" may be a useful predictor of what machines can do, too.[6] Similarly, Andrew Ng, for instance, the former director of the AI Lab at Stanford, looks for tasks that "a typical person can do . . . with less than one second of thought."[7]

The obvious problem with marking out the limits of machines in either of these ways, however, is that any conclusions you reach will become outdated very quickly. Those who try to identify these boundaries are like the proverbial painters of the Forth Rail Bridge in Scotland, a bridge so long they supposedly had to start repainting it as soon as they got to the end—because by then the paint at the other end would have already begun to peel. Spend some time coming up with a sensible account of what machines can do today, and by the time you finish your effort you will probably have to start again and readjust.

A better way to think about machine capabilities is to stop trying to identify specific limits. Repress the temptation to taxonomize, bury the instinct to draw up lists of which particular human faculties are hard to replicate or which particular tasks are intractable, and instead try to make out the more general trends. Do that, and you will see that beneath the particular ripples of progress we see today run some deeper currents.[8] Although it is difficult to say exactly what future machines will be capable of, it is certain that they will be able to do *more* than they can at the moment. Over time, machines will gradually, but relentlessly, advance further into the realm of tasks performed by human beings. Take any technology that currently exists—pick up your smartphone, open your laptop—and you can be confident in saying that this is the *least advanced* that it is ever going to be.

We can think of this general trend, where machines take on more and more tasks that were once performed by people, as "task encroachment."[9] And the best way to see it in action is to look at the three main capabilities that human beings draw on in their work: manual, cognitive, and affective capabilities. Today, each of these is under increasing pressure.

When Daniel Bell, one of the great sociologists of the twentieth century, was pressed to offer some reflections on automation, he quipped that we ought to keep in mind the old Jewish saying: "*for example* is no proof."[10] Yet given the flood of examples to follow, I hope that even Bell would recognize this trend as well.

MANUAL CAPABILITIES

First, take the capabilities of human beings that involve dealing with the physical world, such as performing manual labor and responding to what we see around us. Traditionally, this physical and psychomotor aptitude was put to economic use in agriculture. But over the last few centuries, that sector has become increasingly automated. There are now driverless tractors and cow-milking machines, cattle-herding drones and automated cotton strippers.[11] There are tree-shaking robots that harvest oranges, vine-pruning robots that collect grapes, and vacuum-tube-wielding robots that suck apples off the trees.[12] There are fitness trackers that monitor animal well-being, camera systems that detect unhealthy produce, and autonomous sprayers that drop fertilizer on crops and pesticide on weeds.[13] In Japan, for example, 90 percent of crop spraying is done by unmanned drones.[14] One British farm plants, nurtures, and harvests barley without a person setting foot in the field at all.[15] The US agricultural giant Cargill uses facial recognition software to monitor their cows.[16] The Chinese tech conglomerate Alibaba is developing similar technology to follow pigs, and also plans to use voice recognition software to listen for the squeals of piglets being crushed by their mothers—this, it is thought, will reduce the piglet mortality rate by 3 percent a year.[17]

Today, much of the excitement about automating tasks in the physical world is focused on driverless cars and trucks. In the past, it was thought that the only way for a computer to operate a vehicle was to copy human drivers, to mimic the thinking processes they go through behind the wheel. In keeping with the spirit of the pragmatist revolution, that belief turned out to be wrong: driverless cars, we now realize, do not have to follow fixed, step-by-step rules of the road articulated and set down by human beings, but instead can learn how to navigate by themselves, from the bottom up, drawing on sensor data from millions of real and simulated test drives.[18] Ford has committed to launching a driverless car by 2021.[19] Others have made similar pledges. Tesla claims its cars already have all the hardware needed to drive themselves at a safety level "substantially greater than that of a human driver."[20] Given that, on average, one person is injured in a traffic crash somewhere in the world every second—and one is killed every twenty-five seconds—the prospect of self-driving vehicles is something to be welcomed.[21]

The most immediate impact of driverless vehicles is likely to be on freight delivery, rather than personal transport, due in part to the relative importance of the cargo. A convoy of semiautonomous trucks completed their first trip across Europe in 2016, "platooning" with one another: the front vehicle controlled the speed, and the others automatically followed its lead. (For the moment, there was still a driver in each seat.)[22] And in the future, deliveries may not be done on the road at all. Amazon has filed patents for "drone nests," large beehive-like buildings designed to house fleets of autonomous flying delivery robots, and for "airborne fulfillment centers," airships that cruise at forty-five thousand feet full of products ready for drone delivery.[23]

Airborne robotic delivery might sound fanciful, and Amazon's patents may seem just an attempt to stir up some attention. Yet it is worth remembering that Amazon is among the most advanced users of robotics, with a fleet of over one hundred thousand ground-based robots across its warehouses.[24] And some robots today are already capable of accomplishing remarkable physical feats, like opening doors and climbing walls, ascending stairs and landing backflips, carrying cables over harsh terrain and knotting ropes together in midair.[25] Meanwhile, the global population of industrial robots is rising steadily: the Interna-

Figure 5.1: Global Stock of Industrial Robots (000's)[26]

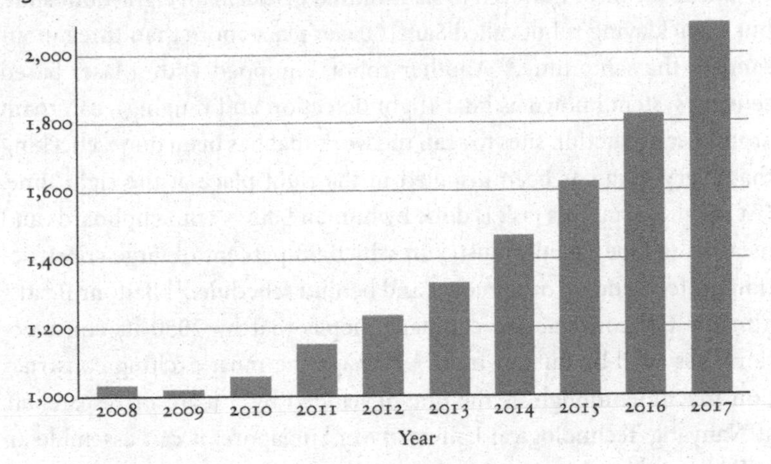

tional Federation of Robotics, a trade association, anticipates there will be more than three million of them in operation in 2020, double the number in 2014.[27]

The car manufacturing industry provides a good case study of task encroachment unfolding in the industrial world. Once upon a time, building a car was a bespoke activity, where craftsmen made each component from scratch. In 1913, Henry Ford automated their craft, replacing handmade components with standardized machine-made parts instead. This allowed him to introduce his famous assembly line, a network of conveyor belts that moved the in-progress cars from worker to worker. Jump forward to today, and robots now account for 80 percent of the work undertaken in manufacturing a car.[28] Nor is this only a story about cars. McKinsey & Company estimate that, as of 2015, 64 percent of worker hours in all areas of manufacturing were spent on tasks that could be automated with *existing* technologies—never mind future ones.[29] (That these activities have not yet been automated, even though it is technically feasible to do so, is an issue we will turn to later in this chapter.)

The construction industry is another part of economic life that has traditionally relied on the manual capabilities of human beings. Now machines are said to be encroaching on tasks here as well. A person may be able to lay three hundred to six hundred bricks in an eight-hour shift, but a bricklaying robot called Sam100 can place more than three thousand in the same time.[30] Another robot, equipped with a laser-based sensing system known as lidar (light detection and ranging), can roam around construction sites to scan the work that has been done, checking that everything has been installed in the right place at the right time. (At the moment, this task is done by human beings with clipboards and measuring tape, in an industry in which 98 percent of large construction projects end up over budget and behind schedule.)[31] Balfour Beatty, a major UK construction company, hopes that by 2050 its construction sites "will be human-free."[32] Perhaps the most exciting construction machine, though, is the one developed by a team of researchers at Nanyang Technological University in Singapore: it can assemble an IKEA chair in twenty minutes.[33]

Builders have also started to adopt 3-D printing techniques, where objects are "printed" out layer by layer, to create entire homes (albeit not very aesthetically pleasing ones at this point). Such 3-D printing technology is not limited to home construction, either: it has been used to print replica motor bikes and edible meals, bikinis and kippot, airplane components and replacement human parts, functional weapons and replicas of sixteenth-century sculptures.[34] General Electric uses 3-D printing to create engine fuel nozzles that are 25 percent lighter and five times more durable than their predecessors. And Médecins Sans Frontières uses it to print prosthetics for the thousands of Syrian refugees who lost limbs in the war, at a fifth of the cost of a conventional replacement.[35]

COGNITIVE CAPABILITIES

Besides manipulating the physical world, machines are also increasingly encroaching on tasks that, until now, have required a human ability to think and reason.

In the legal sphere, for example, JP Morgan has developed a system that reviews commercial loan agreements; it does in a few seconds what would have required, they estimate, about 360,000 hours of human lawyers' time.[36] Likewise, the law firm Allen & Overy has built software that drafts documents for over-the-counter derivatives transactions; a lawyer would take three hours to compile the relevant document, they say, while their system does it in three minutes.[37] A team of American researchers have developed a system that can forecast the outcome of US Supreme Court decisions; it makes correct predictions about 70 percent of the time, whereas human experts, using their legal reasoning, tend to manage only about 60 percent.[38] A team of British researchers have developed a similar system that does the same for the European Court of Human Rights, with 79 percent predictive accuracy.[39] Given that one of the main things that clients involved in a dispute want to know is their chance of winning, systems like these are particularly interesting.

In medicine, many of the most impressive advances have been diagnostic.[40] DeepMind has created a program that can diagnose more than fifty eye diseases with an error rate of only 5.5 percent; when it was compared to eight clinical experts, it performed as well as the two best and outperformed the other six.[41] A team at Oxford has designed a system that can, they claim, outperform cardiologists in anticipating heart problems.[42] In China, where data policy is less restrictive than, say, in the UK or United States, the amount of information reportedly being fed into such systems is astounding. Where the Stanford freckle-analyzing machine, for instance, had 129,450 past cases at its disposal, a diagnostic system created by the Chinese tech giant Tencent and the Guangzhou Second Provincial Central Hospital was able to draw on more than three hundred million medical records from hospitals across the country.[43] Such systems may not be perfectly accurate, but human beings are not infallible, either: today, incorrect diagnoses are said to occur from 10 to 20 percent of the time.[44] That human alternative, not perfection, should be the benchmark for judging the usefulness of these diagnostic machines.

In education, more people signed up for Harvard University's online courses in a single year than had attended the actual university in its entire existence.[45] A big part of my own role at Oxford University is teaching undergraduates economics and mathematics—and alongside my instruction, I often direct my students toward Khan Academy, an online collection of practice problems (100,000 of them, solved two billion times) and instructional videos (5,500 of them, watched 450 million times). Khan Academy has about ten million unique visitors a month, a higher effective attendance than the entire primary- and secondary-school population of England.[46] To be sure, practice problems and online videos, great as they are for making high-quality education content more widely available, are fairly simple technologies. But digital platforms like this are also increasingly being used to support more sophisticated approaches, such as "adaptive" or "personalized" learning systems. These tailor instruction—the content, the approach, and the pace—to the particular needs of each student, in an effort to replicate the one-to-one personal tuition that is provided at a place like Oxford but unaffordable in most other settings. More than seventy companies are developing these systems, and 97 percent of US school districts have invested in them in some form.[47]

The list goes on and on. In finance, computerized trading is now widespread, responsible for about half of all trades on the stock market.[48] In insurance, a Japanese firm called Fukoku Mutual Life Insurance has started using an AI system to calculate policyholder payouts, replacing thirty-four staff in the process.[49] In botany, an algorithm trained on more than 250,000 scans of dried plants was able to identity species in new scans with almost 80 percent accuracy; one paleobotanist, reviewing the results, thought the system "probably out-performs a human taxonomist by quite a bit."[50] In journalism, the Associated Press has begun to use algorithms to compose their sports coverage and earnings reports, now producing about fifteen times as many of the latter as when they relied upon human writers alone. About a third of the content published by Bloomberg News is generated in a similar way.[51] In human resources, 72 percent of job applications "are never seen by human eyes."[52]

We have already seen that machines can now compose music so sophisticated that listeners imagine it must have been written by Bach. There are also now systems that can direct films, cut trailers—and even compose rudimentary political speeches. (As Jamie Susskind puts it, "it's bad enough that politicians frequently sound like soulless robots; now we have soulless robots that sound like politicians."[53]) Dartmouth College, the birthplace of AI, has hosted "Literary Creative Turing Tests": researchers submit systems that can variously write sonnets, limericks, short poems, or children's stories, and the compositions most often taken for human ones are awarded prizes.[54] Systems like this might sound a little playful or speculative; some of them are. Yet researchers who work in the field of "computational creativity" are taking the project of building machines that perform tasks like these very seriously.[55]

At times, the encroachment of machines on tasks that require cognitive capabilities in human beings can be controversial. Consider the military setting: there are now weapons that can select targets and destroy them without relying on human deliberation. This has triggered a set of United Nations meetings to discuss the rise of so-called "killer robots."[56] Or consider the unsettling field of "synthetic media," which takes the notion of tweaking images with Photoshop to a whole new level. There are now systems that can generate believable videos of events that never happened—including explicit pornography that the participants never took part in, or inflammatory speeches by public figures that they never delivered. At a time when political life is increasingly contaminated by fake news, the prospects for the misuse of software like this are troubling.[57]

At other times, the encroachment of machines on tasks that require cognitive capabilities can simply seem peculiar. Take the Temple of Heaven Park in Beijing. In recent years, their public toilets have suffered from a spike in toilet paper thievery. But rather than hire security guards, the park authorities instead installed toilet paper dispensers equipped with facial recognition technology, programmed to provide no more than two feet of paper to the same person in a nine-minute period. The marketing director responsible for the service said the park considered a variety of technological options, but "went with facial recognition, because it's the most hygienic way."[58] Or consider the Catholic Church. In 2011, a bishop issued the first "imprimatur"—the official license granted by church officials to religious texts—known to have been issued to a mobile app. The software is meant to help its users prepare for confession; among its assorted functions, it provides tools for tracking sins and a drop-down menu with various options for acts of contrition. The app caused such a stir that the Vatican itself felt it had to step forward, noting that while people are allowed to use this app to *prepare* for confession, it is not a substitute for the real thing.[59]

AFFECTIVE CAPABILITIES

Beyond the physical world and the cognitive sphere, machines are now also encroaching on tasks that require our affective capabilities, our capacity for feelings and emotions. In fact, an entire field of computer science, known as "affective computing," is dedicated to building systems that do exactly this.

There are systems, for example, that can look at a person's face and tell whether they are happy, confused, surprised, or delighted.[60] Wei Xiaoyong, a Chinese professor at Sichuan University, uses such a program to tell whether his students are bored during class.[61] There are systems that can outperform human beings in distinguishing between a genuine smile and one of social conformity, and in differentiating between a face showing real pain and fake pain. And there are also machines that can do more than just read our facial expressions. They can listen to a conversation between a woman and a child and determine whether they are related, and tell from the way a person walks into a room if they are about to do something nefarious.[62] Another machine can tell whether a person is lying in court with about 90 percent accuracy—whereas human beings manage about 54 percent, only slightly better than what you might expect from a complete guess.[63] Ping An, a Chinese insurance company, uses a system like this to tell whether loan applicants are being dishonest: people are recorded as they answer questions about their income and repayment intentions, and a computer evaluates the video to check whether they are telling the truth.[64]

Then there is the related field of "social robotics." The global population of all robots—not just the industrial ones mentioned before—is now about ten million, with total spending on robotics expected to quadruple from $15 billion in 2010 to $67 billion in 2025.[65] Social robots are a subset of these, distinguished from their mechanical peers by their ability to recognize and react to human emotions. Many of the most striking of these are used in health care. Paro, a therapeutic baby seal, comforts people with cognitive disorders like dementia and Alzheimer's disease; Pepper, a humanoid robot, is used in some Belgian hospitals to greet patients and guide them to the right department.[66] Not everyone gets on well with the machines, though. Pepper, for instance, gained worldwide notoriety in 2015 when a drunk man called Kiichi Ishikawa entered a Japanese cell phone store and attacked the robot that greeted him because "he didn't like its attitude." (The man was promptly arrested.)[67]

These examples of machines that can detect and respond to human emotions are striking. Yet focusing on them too much can mislead us. Why? Because Pepper, Paro, and similar systems are all trying in various ways to mimic the affective capabilities of human beings. The lesson from the pragmatist revolution, though, is that this is not necessary: machines can outperform people at a task without having to copy them.

Think, for instance, about education. It is true that personal contact between a teacher and student is central to the way we educate people today. But that does not stop an online education platform like Khan Academy from providing millions of students every month with high-quality educational material.[68] Likewise, it is true that human interaction between doctors and patients lies at the core of our health care system at the moment. But computer systems do not need to look patients in the eye to make accurate medical diagnoses. And in retail, job ads call for candidates with outstanding "social skills," capable of interacting with customers and persuading them with a smile to open their wallets. But automated checkouts with no social skills at all are replacing friendly cashiers, and online retail involves no human interaction but relentlessly threatens Main Street shops. Roboticists often talk about the "uncanny valley," the observation that when robots become almost—but not perfectly—humanlike in appearance, people feel a sudden discomfort about interacting with them.[69] Yet we may never have to cross this valley at all. It only becomes a problem if our robots are designed to look exactly like human beings—and for most tasks, even affective ones, this is unnecessary.

There is a general lesson here. Economists often label tasks according to the particular capabilities that human beings use to perform them. They talk, for instance, about "manual tasks," "cognitive tasks," and "interpersonal tasks," rather than about "tasks which require manual, cognitive, or interpersonal capabilities *when performed by human beings*." But that way of thinking is likely to lead to an underestimation of quite how far machines can encroach in those areas. As we have seen time and again, machines can increasingly perform various tasks without trying to replicate the particular capabilities that human beings happen to use for them. Labeling tasks according to how humans do them encourages us to mistakenly think that machines could only do them in the same way.

A HEALTHY SKEPTICISM

The list above is not intended to be exhaustive. Some impressive examples are surely missing, while in years to come others will no doubt look tired. Nor are the claims of the various companies mentioned meant to be taken as gospel. At times, it can be hard to distinguish serious corporate ambitions and achievements from mere provocations drawn up by marketers who exaggerate for a living. As Figure 5.2 shows, in mid-2018, "artificial intelligence" and "machine learning" were mentioned about fourteen times as often in the quarterly earnings calls of public companies as they had been just three years earlier. In part, this increase may be driven by genuine technological advances. But it is also likely to be fueled by hype, with companies hastily rebranding old technologies as new AI offerings.

There are some cases of companies selling "pseudo-AIs," chatbots and voice-transcription services that are actually people pretending to be machines (much like the eighteenth-century chess-playing Turk).[70] Less dramatically, but in a similar spirit, a 2019 study found that 40 percent of Europe's AI start-ups actually "do not use any AI programs in their products."[71] There are also notable instances of corporate leaders getting carried away. In 2017, Tesla's CEO Elon Musk expressed his hope that car production in the future will be so highly automated that "air friction" faced by robots would be a significant limiting factor.[72] Just a few months later, under pressure as Tesla failed to meet production targets, he sheepishly tweeted, "yes, excessive automation at Tesla was a mistake."[73]

However, to dwell for too long on any particular omission or exaggeration is to miss the bigger picture: machines are gradually encroaching on more and more tasks that, in the past, had required a rich range of human capabilities. Of course, this has not been a perfectly steady process. Over the years task encroachment has fallen into fallow periods

Figure 5.2: Mentions of "AI" or "Machine Learning" in Earnings Calls[74]

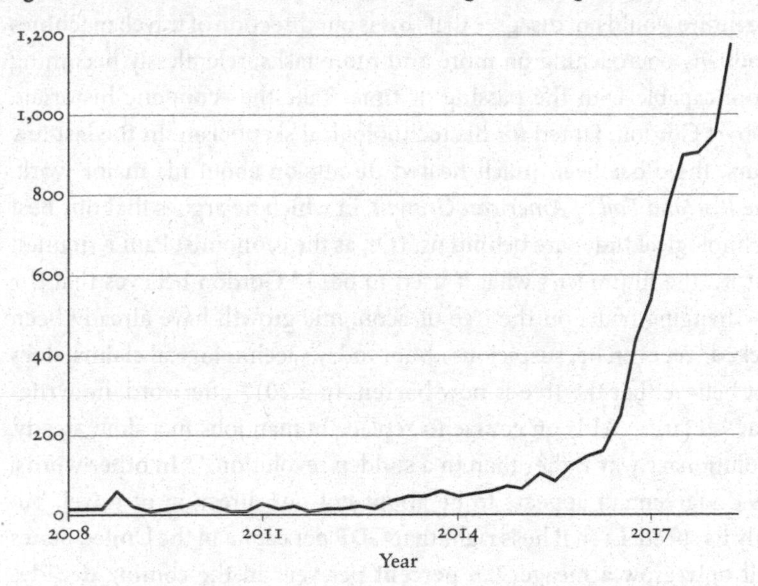

when new obstacles were encountered, surged forward when limits to automation were overcome. Such ebb and flow will surely happen in the future as well. Perhaps new AI winters lie ahead, as today's more feverish enthusiasms about new technologies collide with their actual limitations. But like past limits, many of these will fall away as fresh solutions and work-arounds are fashioned. Economists are wary of labeling any empirical regularity a "rule" or a "law," but task encroachment has proven as law-like as any historical phenomenon can be. Barring catastrophe—nuclear war, perhaps, or widespread environmental collapse—it seems certain to continue.

Isaac Newton famously wrote, "If I have seen further it is by standing on the shoulders of Giants." The same observation applies equally well to the capabilities of machines. Today's technologies build upon those that came before, drawing their strength from the accumulated wisdom of all the past discoveries and breakthroughs. Unless we pack up our creative instincts and put away our impulse to innovate, unless we shout "job done" and wash our hands of AI as a field, the machines we build in the future will be far more capable than they are today.

I hope that even the most conservative economists who think about the future would not disagree that this is our direction of travel: machines gradually encroaching on more and more tasks, relentlessly becoming more capable with the passing of time. Take the economic historian, Robert Gordon, famed for his technological skepticism. In the last few years, there has been much heated discussion about his major work, *The Rise and Fall of American Growth*, in which he argues that our best technological times are behind us. (Or, as the economist Paul Krugman put it, "the future isn't what it used to be.")[75] Gordon believes that the low-hanging fruits on the tree of economic growth have already been picked. Yet even he, suspicious about today's technological claims, does not believe that the tree is now barren. In a 2017 afterword, he writes that "so far . . . AI is on course to replace human jobs in a slow, steady, evolutionary way rather than in a sudden revolution."[76] In other words, his disagreement appears to be about not our direction of travel, but only its speed. Even if he is right that GDP per capita in the United States will only grow a meager 0.8 percent per year in the coming decades (compared to a heady 2.41 percent per year from 1920 to 1970), that still means that in eighty-seven years Americans will be twice as rich as they are now.[77] The debate is only about which generation will enjoy such wealth, not whether future generations will be poorer.

Nevertheless, I still suspect that Gordon's conservatism about the speed of progress is misplaced. Given how popular his view of the future has become, it is worth saying why. Gordon's central claim is that the "special century" of strong economic growth in the United States between 1870 and 1970 "will not be repeated" since "many of the great inventions could happen only once."[78] And, of course, future economic gains will not be driven by the reinvention of electricity, sewage networks, or any other great innovations from the past. The same piece of fruit on the tree cannot be picked twice. But it does not follow that no other great inventions lie in our future. The tree will doubtless bear new fruit in the years to come. *The Rise and Fall of American Growth* is magisterial and yet, in a sense, self-contradictory. It argues with great care that growth was "not a steady process" in the past, yet it concludes that a steady process is exactly what we face in the future: a steady process of *decline* in economic growth, with ever fewer unexpected innovative bursts and technological breakthroughs of the kind that drove our economies onward in the past. Given the scale of investment in technology industries today—many of our finest minds, operating in some of the most prosperous institutions—it seems entirely improbable that there will be no more comparable developments in years to come.

DIFFERENT PACES IN DIFFERENT PLACES

Even though machines are becoming increasingly capable, however, it does not mean that they will be adopted at the same pace in different places around the world. There are three key reasons for this.

Different Tasks

The first reason is the most straightforward: different economies are made up of very different types of jobs, some of which involve tasks that are far harder to automate than others. It is inevitable, therefore, that certain technologies will be far more useful in some places and not others. This is what drives the analysis in Figure 5.3, from the OECD.

Here, "automation risk" is the percentage of jobs in an economy that, in the OECD's opinion, have a greater than 70 percent chance of being automated. As we saw in chapter 2, statements like "job X has a Y percent risk of automation" can be very misleading. Yet this analysis is still useful because it shows how the task composition of different countries can vary wildly. Based on the OECD's judgment about the capabilities of machines, jobs in Slovakia, for instance, are considered to be about six times more at risk of automation than those in Norway. Why? Because jobs in those places involve very different tasks.[79] Figure 5.3 also shows that poorer countries, with lower GDP per head, tend to have a higher "automation risk." The sorts of tasks that the OECD consider to be easiest to automate are disproportionately found in poorer countries. Other research looking explicitly at developing countries has reached the same conclusion as well.[80]

Nor is this only an international story. Even within countries, there can be a huge amount of geographical variation in automation risk. In Canada, for instance, the OECD finds only a 1 percentage point

Figure 5.3: "Automation Risk" v. GDP per Capita[81]

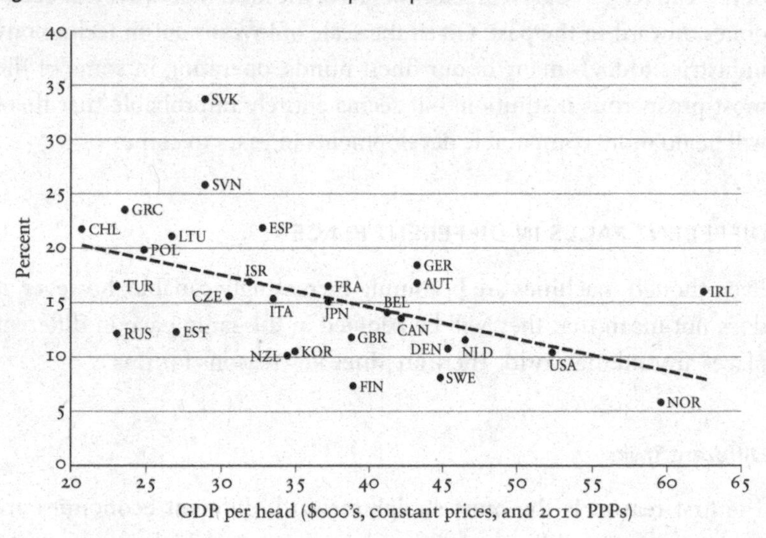

difference between the most at-risk and least at-risk regions, but in Spain that gap is 12 percentage points. Comparing regions can make the international disparities appear to be even more extreme. In West Slovakia, for instance, 39 percent of jobs are said to be at risk of automation, but in the Oslo and Akershus region of Norway only 4 percent are at risk—a more than ninefold difference.[82]

Different Costs

The second reason why machines will be taken up at different paces in different places is cost. Consider this: if you walk through a Moroccan souk today, you may come across craftsmen sitting on the floor and whittling away at pieces of wood using a lathe held between their feet. This is not entirely for show. Because their labor is so cheap, it still makes economic sense for them to continue with their traditional craft rather than turn to any more contemporary woodworking techniques, to use their toes rather than take up any automated tools.[83] The general lesson here is that, in thinking about whether or not it is efficient to use a machine to automate a task, what matters is not only how productive that machine is relative to the human alternative, but also how expensive it is relative to the human alternative. If labor is very cheap in a particular place, it may not make economic sense to use a pricey machine, even if that machine turns out to be very productive indeed.

This sort of reasoning explains why some imagine that low-paid jobs like cleaning, hairdressing, and table-waiting have such a low risk of automation. Not only do they tend to involve "non-routine" tasks, but this work also tends to be lower paid, so the incentive to build machines to take it on is weaker than elsewhere. This is also why the Institute for Fiscal Studies, a leading UK think tank, expressed worries that increasing the minimum wage might increase the risk of automation.[84] If low-paid workers become more expensive, a previously unaffordable machine to replace them might now make financial sense. And this is all the more true for low-paid workers who perform relatively "routine" tasks, such as cashiers and receptionists.

Relative costs can also help explain strange-seeming cases of technological *abandonment*, too. Take the decline of mechanical car washes in the UK. From 2000 to 2015, the number of them installed in roadside garages fell by more than half (from 9,000 to 4,200). Today, the vast majority of car washes in the country are done by hand. Why did automation in the car-washing world go into reverse gear? The Car Wash Association blames immigration, among other factors. In 2004, ten Eastern European countries joined the European Union, and migrants from these countries who arrived in the UK worked for such low wages that they were able to undercut the price of more productive—but also more expensive—mechanical car washes. In this case, cheaper human beings actually managed to displace the machines.[85]

Perhaps the most interesting implications of relative costs are the international ones. In part, these cost variations between countries can explain why new technologies have been adopted so unevenly around the world in the past. A big puzzle in economic history, for instance, is why the Industrial Revolution was *British*, rather than, say, French or German. Robert Allen, an economic historian, thinks relative costs are responsible: at the time, the wages paid to British workers were much higher than elsewhere, while British energy prices were very low. Installing new machines that saved on labor and used readily available cheap fuel thus made economic sense in Britain, whereas it did not in other countries.[86]

More important, relative costs can also explain why new technologies will be adopted unevenly around the world in the future. Take Japan, for example: it is no coincidence that progress in nursing robotics has been particularly swift there. They have one of the largest elderly populations in the world—more than 25 percent are over sixty-five, with the working-age population shriveling at 1 percent a year—and a well-known antipathy toward foreign migrants working in their public services. The result is a shortage of nurses and caregivers (a shortfall expected to reach about 380,000 workers by 2025), and a strong incentive for employers to automate what they do.[87] That is why robots like Paro, the therapeutic robotic seal mentioned before, as well as Robear, which can lift immobile patients from bath to bed, and Palro, a humanoid that can lead a dance class, are being developed and embraced in Japan, whereas elsewhere they are viewed with detached bemusement and disapproval.[88] This story is, in fact, a general one: countries that are aging faster tend to invest more in automation. One study found that a 10 percent increase in the ratio of workers above fifty-six to those between twenty-six and fifty-five was associated with 0.9 more robots per thousand workers. In 2014, there were only 9.14 industrial robots per thousand workers in US manufacturing, far lower than, for instance, in Germany, which had 16.95 robots per thousand workers—but if the United States had the same demographics as Germany, this study implies, the difference would have been 25 percent smaller.[89]

Still, while countries and regions and particular parts of the economy may vary in relative costs, they are all being pulled in the same direction. New technologies are not simply becoming more powerful in various settings but, in many cases, more affordable as well. Consider the cost of computation: as shown in Figure 5.4, it plummeted over the second half of the twentieth century, in a mirror image of the explosion in computational power that took place during this time. (Here, again, the *y*-axis has a logarithmic scale, with one step down representing a tenfold decrease in cost, two steps representing a hundredfold decrease, and so on.)

Michael Spence, a Nobel laureate in economics, estimates that the

Figure 5.4: The Cost per Million Computations, 1850–2006 (2006 $'s)[90]

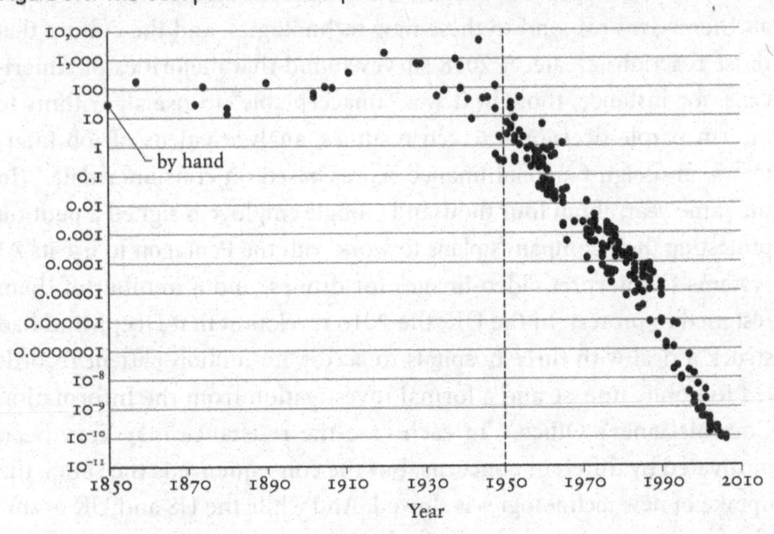

cost of processing power fell by a factor of roughly ten billion times in the last fifty years of the twentieth century.[91] Trends as powerful and persistent as this eventually catch up with virtually all places and parts of economic life, however much the relative costs may differ.

Different Regulations, Contrasting Cultures

The final reason why the machines will be taken up at different rates in different places has to do not with economics, but with the regulatory and cultural environments in which new technologies are deployed. The regulatory side of this is continually changing: over the last few years, for instance, almost all developed countries have published some form of "AI strategy," setting out how they hope to shape the field. China has published a plan to become "the front-runner" in AI by 2030, demanding that research take place "everywhere and at every moment." President Vladimir Putin has declared that "whoever becomes the leader in this sphere [AI] will become the ruler of the world."[92] Ambitions like this mean that the formal regulatory environments in which these new technologies are developed and put to use are attracting more and more attention.

Equally as important as state intentions, though, is how individuals themselves respond to these new technologies, and the culture that those reactions create. A 2018 survey found that majorities of Americans, for instance, thought it was "unacceptable" to use algorithms to inform parole decisions, screen résumés, analyze videos of job interviews, or assign personal finance scores based on consumer data.[93] In the same year, about four thousand Google employees signed a petition protesting their company's plans to work with the Pentagon to use its AI systems to interpret video images for drones, and a number of them resigned in protest. In the UK, the 2016 revelation that DeepMind had struck a deal with three hospitals to access 1.6 million patient records led to public unease and a formal investigation from the Information Commissioner's Office.[94] In each case the resistance may have been motivated by different concerns, but the consequence is the same: the uptake of new technologies is slowed. And while the US and UK examples here are sensitive ones, it is also important to keep in mind that even the most innocuous technologies can still attract cultural conservatism with a similar effect. It took two decades, for instance, for the stethoscope to start being routinely used by doctors after its invention in 1816. The medical men, it is said, did not want an instrument to "get between their healing hands and the patient."[95]

The Story of China

To see all of these factors at work in one place, consider China. Its remarkable growth over the last few decades has been, in large part, powered by cheap workers who used to be in agriculture. Lots of them, in search of better pay, were drawn in by the bright lights of factories in increasingly prosperous cities. For a time, employing these workers rather than machines made economic sense because their wages were so low. However, change is afoot. First, the Chinese economy appears to be composed of tasks that are particularly susceptible to automation: researchers claim that 77 percent of jobs there are "at risk."[96] Second, relative costs are rising: from 2005 to 2016, wages trebled.[97] This means it now makes greater financial sense to replace human beings with machines. And third, the regulatory and cultural environment is supportive of doing so. As early as 2014, President Xi Jinping called for a "robot revolution," and any resistance to that ambition from civil society is unlikely to be met as accommodatingly as in the West.[98]

Together, this explains why more robots were installed in China in 2016 than in any other country: almost a third of all the robots installed worldwide, and more than twice the number in South Korea, which took a distant second place.[99] And the country is making great strides in AI research, as well. At the inaugural 1980 meeting of the Association for the Advancement of Artificial Intelligence, a key conference for the field, there were no papers from Chinese researchers; most were written by those in the United States. In 1988, there was one Chinese submission, as the United States again dominated. But in 2018, China submitted 25 percent more papers to the conference than the United States, and was only three papers behind in acceptances.[100] Today, if we look at the top 1 percent of the most highly cited papers in mathematics and computer science, the two universities producing the greatest number of such papers are both in China, ahead of Stanford and MIT.[101]

Again, though, the different paces in different places matter less than the general trend. Almost everywhere, machines are becoming increasingly capable, creeping ever further into the realm of tasks once performed only by human beings. To adapt the old saying, nothing in life can be said to be certain, except death, taxes—and this relentless process of task encroachment.

6

Frictional Technological Unemployment

When John Maynard Keynes popularized the term "technological unemployment" about ninety years ago, he prophesied that we would "hear a great deal in the years to come" about it.[1] Despite his clarity about the threat, however, and his prescience about the anxiety that would accompany it, he did not really explain how technological unemployment would happen. He described it as arising "due to our discovery of means of economising the use of labour outrunning the pace at which we can find new uses for labour," but offered very few specifics. Instead, he merely gestured to the "revolutionary technical changes" taking place to persuade his readers to agree with his argument. And nod they probably did: Keynes was writing in the wake of the Roaring Twenties, a time that saw remarkable technological progress in everything from airplanes to antibiotics to "talkies" in movie theaters. It would turn out, though, that the nature and scale of these innovations were not a reliable guide to what lay ahead. Nor is Keynes's definition of technological unemployment particularly revealing. It leaves open the most important question: *Why* might we not find new uses for human labor in the future?

As we discussed, the future of work depends on two forces: a harmful substituting force and a helpful complementing one. Many tales have a hero and a villain fighting each other for dominance, but in our story, technology plays both roles at once, displacing workers while simultaneously raising the demand for their efforts elsewhere in the economy. This interaction helps explain why past worries about automation were misplaced: our ancestors had predicted the wrong winner in that fight, underestimating quite how powerful the complementing force would prove to be or simply ignoring that factor altogether. It also helps to explain why economists have traditionally been dismissive of the idea of technological unemployment: there appeared to be firm limits to the substituting force, leaving lots of tasks that could not be performed by machines, and a growing demand for human beings to do them instead.

But economists' dismissal of technological unemployment is misconceived. The pragmatist revolution has shown that those supposedly firm limits to the capabilities of machines are not so firm after all. The substituting force is gathering strength, and new technologies are not politely following the boundaries that some forecasters had set down, marking out what tasks they can and cannot do. Of course, this is not necessarily a problem. Economic history shows that as long as the complementing force is strong enough, it does not matter if machines can substitute for human beings at a wider range of tasks—there will still be demand for human work in other activities. We still live in an Age of Labor, as we have since the Industrial Revolution began.

The challenge for those of us who agree with Keynes, then, is to explain how technological unemployment might be possible, without neglecting—as people have done in the past—the helpful complementing force.

WORK, OUT OF REACH

Greek mythology tells of a man called Tantalus who commits an unsavory crime: chopping up his own son and serving him as a meal to the gods. Given his dinner guests' omniscience, this turns out to be a very bad decision. When he is found out, his punishment is to stand, for eternity, in a pool of water up to his chin, surrounded by trees bursting with fruit—but the water recedes from his lips whenever he tries to take a sip, and the tree branches swing away when he reaches out to pick their bounty.[2] The story of Tantalus, which gives us the word *tantalize*, captures the spirit of one kind of technological unemployment, which we can think of as "frictional" technological unemployment. In this situation, there is still work to be done by human beings: the problem is that not all workers are able to reach out and take it up.[3]

Frictional technological unemployment does not necessarily mean there will be fewer jobs for human beings to do. For the next decade or so, in almost all economies, the substituting force that displaces workers is likely to be overwhelmed by the complementing force that raises the demand for their work elsewhere. Despite all the technological accomplishments we have seen in recent decades, vast areas of human activity cannot yet be automated, and limits to task encroachment still remain in place. The historical trend—where there is always significant demand for the work of human beings—is likely to go on for a while. But as time passes, this will be of comfort to a shrinking group of people. Yes, many tasks are likely to remain beyond the capabilities of machines, and technological progress will tend to raise the demand for human beings to do them. However, as in the tale of Tantalus, this in-demand work is likely to be agonizingly out of the grasp of many people who want it as well.

"Frictions" in the labor market prevent workers from moving freely into whatever jobs might be available. (If we think of the economy as a big machine, it is as if there is sand or grit caught up in its wheels, stopping its smooth running.) Today, there are already places where this is happening. Take men of working age in the United States, for instance. Since World War II, their participation in the labor market has collapsed: one in six are now out of work, more than double the rate of 1940.[4] What happened to them? The most compelling answer is that these men fell into frictional technological unemployment. In the past, many of them would have found well-paid work in the manufacturing sector. Yet technological progress means that this sector no longer provides sufficient work for them all to do: in 1950, manufacturing employed about one in three Americans, but today it employs fewer than one in ten.[5] Plenty of new jobs have been created in other sectors as the US economy changed and grew—since 1950, it has expanded about fourfold—but critically, many of these displaced men were not able to take up that work. For a variety of reasons, it lay out of their reach.[6]

In the coming decade, this is likely to happen to other types of workers as well. Like those displaced manufacturing workers, they, too, will become trapped in particular corners of the labor market, unable to take up available work elsewhere. There are three distinct reasons for that, three different types of friction at work: a mismatch of skills, a mismatch of identity, and a mismatch of place.

THE SKILLS MISMATCH

In many developed economies, as we have seen, the labor market has become increasingly polarized in recent years. There is more high-paid, high-skill work at the top than there used to be, and plenty of low-paid or low-skilled work at the bottom, but the jobs between these—which traditionally supported many people in well-paid middle-class employment—are withering away. If the United States is at all representative, the evidence suggests that for now this hollowing out is likely to continue.[7] And from it comes the first reason to expect frictional technological unemployment: the leap to the top is increasingly difficult to make.

In the past, it was possible to ride successive waves of technological progress up through the labor market. A few hundred years ago, when machines drove human beings from a traditional life on the fields, those people transitioned into manufacturing with relative ease. The shift from farms to factories meant that work changed, but the new skills required were readily attainable; it was still a form of manual work. Then, as the Industrial Revolution gathered pace, machines became more complex, production processes more sophisticated, and factories more unwieldy. Demand rose for better-educated blue-collar workers—engineers, machinists, electricians, and the like—and for white-collar workers to manage operations and provide professional services. This move from manual work toward cognitive work was more challenging for workers who wanted to be upwardly mobile. As Ryan Avent, a senior editor at the *Economist*, notes, few people in the early nineteenth century would have been well-prepared for it: "most were illiterate and innumerate."[8]

Nevertheless, for many people it was still possible to learn the right skills. And as a newfound enthusiasm for mass education took root in the late nineteenth and early twentieth centuries, that, too, helped to sweep many people up and along.

As the twentieth century rolled on, skill levels continued to rise around the world as people tried to clamber up into better-paid work. Economists spoke of a metaphorical race that was taking place between workers and technology, implying that people simply had to learn the right skills to keep up.[9] But today, this race looks like a far tougher undertaking for those who want to take part.

For one thing, many people in the race are already running as fast as they can. Around the world, the proportion of people getting a good education has stopped growing. As Avent points out, it is very difficult to get more than 90 percent of people to finish secondary school, or more than 50 percent to graduate from college.[10] Research by the OECD found a similar plateauing in workers' skill levels as well. "Most countries around the world have worked to increase the education and skills of their populations," the OECD reports. "However, the available data on adult skills in OECD countries over the past two decades does not show a general increase in the proportion of workers at higher proficiency levels as a result of past education improvements."[11]

At the same time, human beings are likely to find this race with technology ever harder, because its pace is accelerating. Literacy and numeracy are no longer enough to keep up, as they were when workers first made the move from factories to offices at the turn of the twentieth century. Ever higher qualifications are required. Notably, while workers with a college degree have been outperforming those with only a high school education, those with postgraduate qualifications have seen their wages soar far more, as seen in Figure 2.3 on page 33.[12]

The accelerating pace of the race explains, in part, why Silicon Valley responded to President Donald Trump's immigration controls with such emphatic, collective disapproval. As part of his "America First" policy, Trump promised to restrict the "specialty occupation" H-1B visas that allow around eighty-five thousand foreigners into the United States each year, often to work at high-tech companies. Silicon Valley has a significant appetite for high-skilled workers and relies upon these visas to bring in foreign workers to satisfy this demand. The idea is that Americans are not always up to the job: companies only apply for visas, they say, when they cannot find qualified people for them at home.[13] We might be slightly suspicious of that claim; critics say that companies actually use these visas to employ foreign workers whom they can pay lower wages.[14] Still, it is estimated that there are only twenty-two thousand PhD-educated researchers in the world who are capable of working at the cutting edge of AI, and only half are based in the United States—a large proportion, but still a comparatively small number of possible workers given the sector's importance.[15]

THE IDENTITY MISMATCH

For those who are unable to reach high-paid, high-skilled work, an inevitable alternative is a retreat to less-skilled or lower-paid work. This certainly appears to have been the fate of less-educated workers in the United States: they have moved, in David Autor's words, "less and less upwards" through the labor market.[16]

Strikingly, though, over the last decade and a half, many well-educated people aiming for the top of the labor market have also missed it and been forced to move down into work for which they are overqualified. Fast-food jobs, for instance, were seen in the 1950s and '60s as being largely for "teens in summertime," but in the United States today only a third of fast-food workers are teenagers, 40 percent are older than twenty-five, and nearly a third have some college education.[17] More broadly, a third of Americans with degrees in STEM subjects (science, technology, engineering, and math) are now in roles that do not require those qualifications.[18] And when economists took all the jobs performed by US college graduates and examined the tasks that make them up, they found a collapse in the "cognitive task intensity" of these roles from 2000 onward—a "great reversal in the demand for skills."[19] As shown in Figure 6.1, graduates are increasingly finding themselves in roles that are less cognitively demanding and less skilled than before.

Not everyone has retreated in this way, though. Many people have rejected this movement into lower-paid or lower-skilled roles, choosing to fall into unemployment instead. And this is the second reason

Figure 6.1: Cognitive Task Intensity of Jobs Performed by College Graduates (Index 1980 = 1)

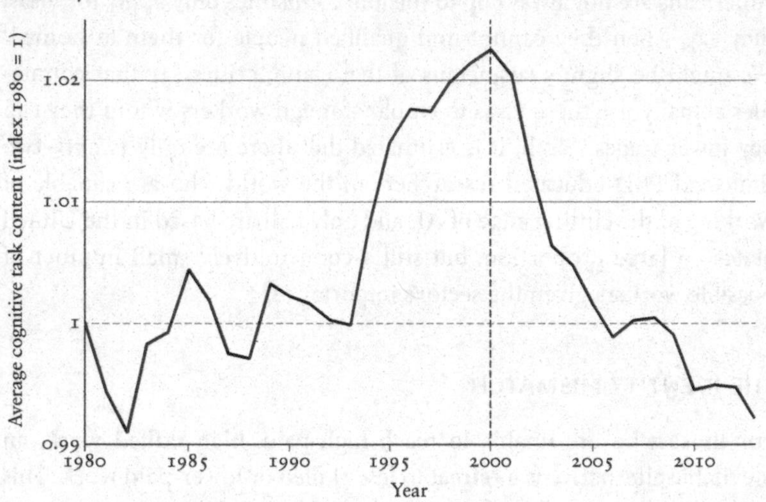

to expect frictional technological unemployment in the future. People might not only lack the skills to do more and more of the available work, but they might also be unwilling to do lower-skilled work that is on offer.

In South Korea, something like this is already happening. It is a country famed for the intensity of its academic culture, where about 70 percent of young people have degrees. But half of the unemployed there are college graduates as well.[20] In part, this is because these highly qualified people are reluctant to take up the work that is available to them—poorly paid, insecure, or low-status roles, simply not what they imagined they were training to become.[21]

The fact that workers are willing to shun employment like this is particularly important because there is no reason to think that technological progress will necessarily create more appealing work in the future. There is a common fantasy that technological progress must make work more interesting—that machines will take on the unfulfilling, boring, dull tasks, leaving behind only meaningful things for people to do. They will free us up, it is often said, to "do what really makes us human." (The thought is fossilized in the very language we use to talk about automation: the word *robot* comes from the Czech *robota*, meaning drudgery or toil.) But this is a misconception. We can already see that a lot of the tasks that technological progress has left for human beings to do today are the "non-routine" ones clustered in poorly paid roles at the bottom of the labor market, bearing little resemblance to the sorts of fulfilling activities that many imagined as being untouched by automation. There is no reason to think the future will be any different.

For adult men in the United States, a similar story is unfolding, where some workers likewise appear to have left the labor market out of choice rather than by necessity—though for a different reason. Displaced from manufacturing roles by new technologies, they prefer not to work at all rather than take up "pink-collar" work—an unfortunate term intended to capture the fact that many of the roles currently out of reach of machines are disproportionately held by women, like teaching (97.7 percent of preschool and kindergarten teachers are women), nursing (92.2 percent), hairdressing (92.6 percent), housekeeping (88 percent), social work (82.5 percent), and table-waiting (69.9 percent).[22] While male-dominated roles in production are on the decline, such female-dominated roles are on the rise and expected to create the most jobs in the coming years, as shown in the projections from the US Bureau of Labor in Figure 6.2.[23]

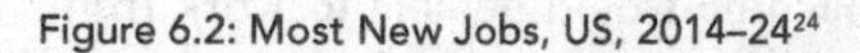
Figure 6.2: Most New Jobs, US, 2014–24[24]

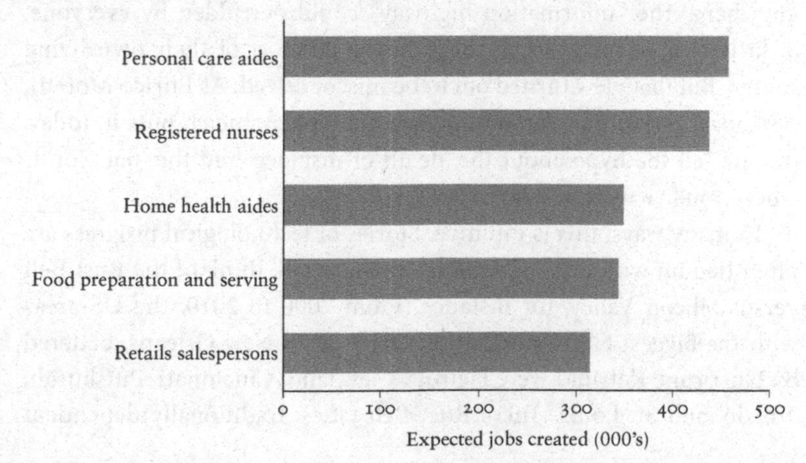

Why are people reluctant to take on available work that they are capable of doing? It doesn't help that the pay for most of these "pink-collar" jobs is significantly below the national average.[25] But even more important, it seems, many of the male workers are attached to an identity that is rooted in a particular sort of role—its social status, the nature of the work, the type of people that tend to do it—and are willing to stay unemployed in order to protect that identity.[26]

THE PLACE MISMATCH

The third cause of frictional technological unemployment is that the work that exists may simply be in the wrong geographical area. People may have the skills to take it on and the appetite to do it—yet they still cannot relocate to take it up. There are many reasons why a move might be out of the question. Some workers just don't have the money that it takes to pick up and move. Others do not want to leave a community or a place that is important to them. In either case, the consequences are the same: technological change might stir up the demand for the work of human beings, but not in the particular place a person happens to be located.

Back in the early days of the Internet, there was a moment when it seemed that these sorts of worries about location might no longer matter. Anybody with a connected machine would be able to work from anywhere. The "information highway" could be ridden by everyone, at little cost or inconvenience, from the comfort of their own living rooms. But that view turned out to be misconceived. As Enrico Moretti, perhaps the leading scholar studying these phenomena, puts it, today, despite "all the hype about the 'death of distance' and the 'flat world,' where you live matters more than ever."[27]

In many ways, this is intuitive. Stories of technological progress are often tied up with tales of regional rise and fall: think of the Rust Belt versus Silicon Valley, for instance. From 2000 to 2010, the US areas with the biggest fall in population (aside from New Orleans, battered by Hurricane Katrina) were Detroit, Cleveland, Cincinnati, Pittsburgh, Toledo, and St. Louis. These Rust Belt cities, traditionally dependent upon manufacturing, lost up to 25 percent of their population as that sector was pushed into decline.[28] In Silicon Valley, meanwhile, technological advances were responsible for stratospheric takeoff. Today, it is the innovation capital of the world, producing more patents, creating more high-tech jobs, and attracting more venture capital funding than anywhere else in the country by a long shot.[29]

Nor is it the case that Silicon Valley and places like it only produce the sort of high-skilled work that we tend to associate with large technology companies. Yes, it is true that, if you are a computer scientist, you are going to find the best-paid work in Silicon Valley: in places like Boston, New York, or Washington, DC, you might earn up to 40 percent less. But even if you are not high-skilled in that way, being near places like Silicon Valley has tended to be worthwhile in the past. San Francisco and San Jose, for instance, have some of the best-paid barbers and waiters in the United States, alongside many other roles.[30] In fact, dense urban areas of all types have traditionally meant higher wages for everyone there, skilled or unskilled. (Whether this will continue, though, is not clear: according to the most recent data, this is still the case for well-educated workers, but for the less-educated the benefits of being in a big city have "mostly disappeared.")[31]

If you look at US statistics, you might be tempted to think that the threat of place mismatch is overstated. On the face of it, Americans appear to be remarkably mobile: about half of households change their address every five years, and the proportion of people living in a different state from the one where they were born has risen to one-third.[32] But there are two important caveats. First, this is not the case everywhere. Europeans, for instance, are far more immovable: 88.3 percent of Italian men aged between sixteen and twenty-nine still live at home.[33] And second, those who do move tend to be better educated as well. In the United States, almost half of college graduates move out of their birth states by the time they are thirty, but only 17 percent of high school dropouts do so.[34] People may be pouring out of places like Detroit but, as Moretti puts it, "the flow of high school dropouts is a mere trickle." In addition to all of the other inequalities in the economy, there is a large "inequality of mobility" as well.

NOT JUST UNEMPLOYMENT

One of the unhelpful consequences of the term "technological unemployment" is that it encourages us to think that the only (or at least the main) way that new technologies will affect the world of work is by changing the unemployment rate—the percentage of workers in the labor market who are looking for a job and cannot find one. That figure alone, though, does not capture the full picture. To begin with, some people, facing the mismatches of skills, identity, and place, might simply give up on the job hunt and drop out of the labor market altogether. If that were to happen, the official unemployment rate would actually *fall*: since those people were no longer searching for work, they would not count as being unemployed for the purposes of that statistic.

It is important, then, to also pay attention to what is known as the "participation rate": the percentage of people in the entire working-age population (not just those actively in the labor market) who are employed. In the United States today, for instance, the unemployment rate is an impressively low 3.7 percent. At the same time, however, the participation rate has collapsed, falling to its lowest level since 1977. More and more working-age Americans, it appears, are abandoning the world of work altogether—and that should be a cause for alarm.[35] Similarly, in the future, we should be cautious about focusing exclusively on the unemployment rate, and keep an eye on the participation rate as well.

The more fundamental problem with the unemployment rate, though, is that it is only concerned with the number of available jobs, not their nature. Economic history clearly shows, however, that new technologies may reduce not only the amount of work to be done, but also the attractiveness of that work. The Industrial Revolution gave us a glimpse of that: there were no vast pools of displaced people, but much of the work that emerged was not particularly pleasant. This will be true in the future as well.

In a sense, this is obvious. Some workers, rather than dropping out of the labor market because they lack the right skills, dislike the available jobs, or live in the wrong place, will instead pursue whatever work does remain for them to do. And when this happens—where workers find themselves stranded in a particular corner of the labor market but still want a job—the outcome will not be technological unemployment, with people unable to find work at all, but a sort of technological overcrowding, with people packing into a residual pool of whatever work remains within their reach. Rather than directly cause a rise in joblessness, this could have three harmful effects on the nature of the work.

The first is that, as people crowd in, there will be downward pressure on *wages*. Curiously, whereas technological unemployment is a controversial idea in economics, such downward pressure is widely accepted.[36] At times it can be puzzling that economists tend to make such a hard distinction between no work and lower-paid work. The two are treated as unrelated phenomena—the former regarded as impossible, the latter as entirely plausible. In practice, the relationship between the two is far less straightforward. It seems reasonable to think that as more people jostle for whatever work remains for them to do, wages will fall. It also seems reasonable to think that these wages might fall so low in whatever corner of the labor market a worker is confined to that it will no longer be worth their while to take up that work at all. If that happens, the two phenomena become one. This is not an unlikely possibility: in 2016, 7.6 million Americans—about 5 percent of the US workforce—who spent at least twenty-seven weeks of the year in the labor force still remained below the poverty line.[37] The same thing could happen elsewhere in the economy, too.

The second impact of people crowding into the work that remains is that there will be downward pressure on the *quality* of some of the jobs as well. With more workers chasing after those jobs, there is less need for employers to attract them with good working conditions. Karl Marx spoke of workers as the "proletariat," adopting the ancient Roman term for members of the lowest social class; today, though, the term *precariat* is gaining ground instead—a word that captures the fact that more and more work is not just poorly paid, but also unstable and stressful.[38] It is sometimes said, in a positive spirit, that new technologies make it easier for people to work flexibly, to start up businesses, become self-employed, and to have a more varied career than their parents or grandparents. That may be true. But for many, this "flexibility" feels more like instability. A third of the people who are on temporary contracts in the

UK, for instance, would prefer a permanent arrangement; almost half on zero-hour contracts want more regular work and job security.[39]

The third impact of people crowding in on the work that remains involves the *status* attached to it. James Meade, a Nobel laureate in economics, anticipated this when he wrote about the future of work in 1964. He imagined that many future jobs would, in different ways, involve providing certain types of support to the more prosperous. He thought that the future would be made up of "an immiserized proletariat and of butlers, footmen, kitchen maids, and other hangers-on"—and, in a certain sense, he was right.[40] Parts of our economic life already feel two-tiered in the way that Meade imagined: many of those fast-growing jobs in Figure 6.2, for instance, from retail sales to restaurant serving, involve the provision of low-paid services to the wealthy. But these "hangers-on" need not be all be "immiserated," as Meade expected. In rich corners of cities like London and New York it is possible to find odd economic ecosystems full of strange but reasonably well-paid roles that rely almost entirely on the patronage of the most prosperous in society: bespoke spoon carvers and children's playdate consultants, elite personal trainers and star yoga instructors, craft chocolatiers and artisanal cheesemakers. The economist Tyler Cowen put it well when he imagined that "making high earners feel better in just about every part of their lives will be a major source of job growth in the future."[41] What is emerging is not just an economic division, where some earn much more than others, but a status division as well, between those who are rich and those who serve them.

To see how frictional technological unemployment might play out in the future, consider the millions of people who make their living as drivers in the United States. In a world of driverless vehicles, many of those jobs—between 2.2 and 3.1 million of them, if President Obama's White House is to be believed—will be wiped out.[42] At the same time, new work may appear elsewhere. Perhaps there will be a surge in demand for computer scientists, who are capable of designing, calibrating, and maintaining the driverless fleets. Or perhaps a more prosperous economy will lead to greater demand for unrelated low-skill services, like cleaning and hairdressing and gardening. But the out-of-work drivers may not be well placed to take up any of these new opportunities. It is not so easy for truckers to retrain as programmers. They may dislike the character of these new roles. And even if they wanted to take one up, they may not live in the right place to do so. In this way, skills mismatch, identity mismatch, and place mismatch might affect the displaced workers all at the same time.

The idea of frictional technological unemployment may not resemble some of the more dramatic imagery associated with the future of work. Some people may question whether this is "real" technological unemployment, because if workers learned the right skills, changed the way they think about themselves, or simply moved to where work is, the friction would disappear. But it would be a mistake to dismiss the problem on these grounds. While in theory it may be only an issue temporarily, in practice such frictions are very hard to resolve. And from the point of view of the workers, there is not a meaningful distinction between work that's out of their grasp and no work at all. For them, tales of islands of employment elsewhere in the economy might as well be fairy tales.

7

Structural Technological Unemployment

A few years ago, Chris Hughes, one of Facebook's cofounders, was at a dinner with a crowd of influential economists and high-powered policymakers. As part of the evening, Jason Furman, the chair of President Obama's Council of Economic Advisers at that time, was invited to deliver a presentation to the assembled guests on the topic of "digital competitiveness." Hughes, who was interested in the future of work, interrupted him halfway through to ask: "What are you doing to plan for a future with more artificial intelligence where there might be fewer jobs?" Furman responded—"barely concealing his annoyance," according to Hughes—that "three hundred years of history tells us that can't be true."[1]

In my experience, economists tend to be comfortable with the idea of "frictional" technological unemployment, the type we explored in the previous chapter. They can readily picture a future where there is lots of work to be done but some people cannot do it. Hughes, though, was asking Furman about a different problem. He wanted to know what was being done to prepare for a future that did not have enough work in it for human beings—full stop. We can think of this kind of scenario, in which there are actually too few jobs to go around, as "structural" technological unemployment. Most economists, like Furman, tend to be far less willing to accept this as a possibility.[2]

Are they right? Does the fact that after three centuries of radical technological change there is still enough work for people to do tell us that there will *always* be sufficient demand for the work of human beings? I do not think so. Yes, history may tell us that in the past there has been enough demand to keep nearly everyone employed. But it does not guarantee that this must also be true in decades to come. Until now, the substituting force that displaces workers has been weaker than the complementing force that raises demand for their work elsewhere. But it is likely that this balance between the forces will tip the other way in the future—and tip that way permanently.

A WEAKENING COMPLEMENTING FORCE

There is little doubt that as task encroachment continues, with machines taking on more and more tasks, the harmful substituting force will grow stronger. Workers will be displaced from a wider range of activities than ever before. Why, though, can we not simply rely on the complementing force to overcome that effect, as it has done until now? Why would it fail to act as a bulwark against the substituting force? The answer is that task encroachment also has a second pernicious effect: over time, it is likely not just to strengthen the substituting force, but to wear down the complementing force as well.

In the past, as we have seen, the complementing force raised the demand for displaced workers in three ways: through the productivity effect, the bigger-pie effect, and the changing-pie effect. Together, they ensured there was always enough work for people to do. But in the future, as machines continue their relentless advance, each of these effects is likely to be drained of its strength.

The Productivity Effect

The first way that the complementing force has worked so far is through the productivity effect. Machines displaced people from certain tasks, but they also made workers more productive at other activities, ones that were not being automated. When these improvements in worker productivity were passed on to consumers (through lower prices or higher-quality offerings), that helped to raise the demand for those workers' efforts.

In the future, new technologies will no doubt continue to make some people more productive at certain tasks. But this will only continue to raise the demand for human workers if they remain better placed to do those tasks than a machine. When that ceases to be the case, improvements in worker productivity will become increasingly irrelevant: machines will simply take their place instead.[3]

Think of a traditional craft like candle making or cotton spinning. Human beings were once best-placed to do these jobs. Today, though, they are almost entirely done by machines. Perhaps there are some hobbyists who are still interested in how well human beings can perform the tasks involved—in how many candles a present-day tallow-chandler could make or how much cotton thread a contemporary weaver could spin using modern tools. Yet from an economic point of view, those human capabilities no longer matter at all. It is more efficient to just automate all of these activities.

As task encroachment continues, human capabilities will become irrelevant in this fashion for more and more tasks. Take sat-nav systems. Today these make it easier for taxi drivers to navigate unfamiliar roads, making them better at the wheel. At the moment, therefore, they complement human beings. But this will only be true as long as human beings are better placed than machines to steer a vehicle from A to B. In the coming years, this will no longer the case: eventually, software is likely to drive cars more efficiently and safely than human beings can. At that point, it will no longer matter how good people are at driving: for commercial purposes, that ability will be as amusingly quaint as our productivity at hand-fashioning candles or cotton thread.[4]

Chess provides another illustration of how the productivity effect will fade away in the years to come. For some time, Garry Kasparov has celebrated a phenomenon that he calls "centaur chess," which involves a human player and a chess-playing machine working together as a team. Kasparov's thought was that such a combination would beat any chess computer playing alone.[5] This is the productivity effect in action: new technologies making human beings better at what they do. The problem, though, is that Kasparov's centaur has now been decapitated.

In 2017, Google took AlphaGo Zero, the go-playing machine that trains itself, tweaked it so it could play other board games as well, and gave it the rules of chess. They called the new system AlphaZero. Instead of absorbing the lessons of past games by the best human chess players, this machine had no human input at all. Yet after only a day of self-training, it was able to achieve unparalleled performance, beating the best existing chess-playing computer in a hundred-game match—without losing a single game.[6] After that trouncing, it is hard to see what role human players might have alongside a machine like this. As Tyler Cowen put it, "the human now adds absolutely nothing to man-machine chess-playing teams."[7]

There is a deeper lesson here. Kasparov's experiences in chess led him to declare that "human plus machine" partnerships are the winning formula not only in chess, but across the entire economy.[8] This is a view held by many others as well. But AlphaZero's victory shows that this is wrong. Human plus machine is stronger only as long as the machine in any partnership cannot do whatever it is that the human being brings to the table. But as machines become more capable, the range of contributions made by human beings diminishes, until partnerships like these eventually just dissolve. The "human" in "human plus machine" becomes redundant.

The Bigger-Pie Effect

The second way that the complementing force has helped human beings is through the bigger-pie effect. If we think of a country's economy as a pie, technological progress around the world has made virtually all of those pies far, far bigger. This has meant that workers who found themselves displaced from one part of the economy could find work in another part of it instead, as growing incomes led to increased demand for their efforts elsewhere.

In the future, economic pies will no doubt continue to grow, incomes will be larger than they have ever been, and demand for goods will soar. Yet we cannot rely on this to necessarily bolster the demand for the work of human beings, as it has in the past. Why? Because just as with the productivity effect, the bigger-pie effect will only help if people, rather than machines, remain better placed to perform whatever tasks have to be done to produce those goods.

For now, that may be a reasonable expectation. We live in the Age of Labor, and if new tasks have to be done it is likely that human beings will be better placed to do them. But as task encroachment continues, it becomes more and more likely that a machine will be better placed instead. And as that happens, a growing demand for goods may mean not more demand for the work of human beings, but merely more demand for machines.

We can already catch a glimpse of this phenomenon at work. Take the UK agricultural sector, for example. This part of the British economic pie has grown dramatically over the last century and a half, yet that has not created more work for people to do in it. Today, British agriculture produces more than five times the output that it did in 1861, but the proportion of the total UK workforce employed in it has fallen from 26.9 percent to 1.2 percent, and the number of actual workers in the sector has shrunk almost tenfold, from 3.2 million to 380,000. More money is spent on agricultural output than ever before,

Figure 7.1: UK Agriculture, 1861–2016 (Index 1861 = 100)[9]

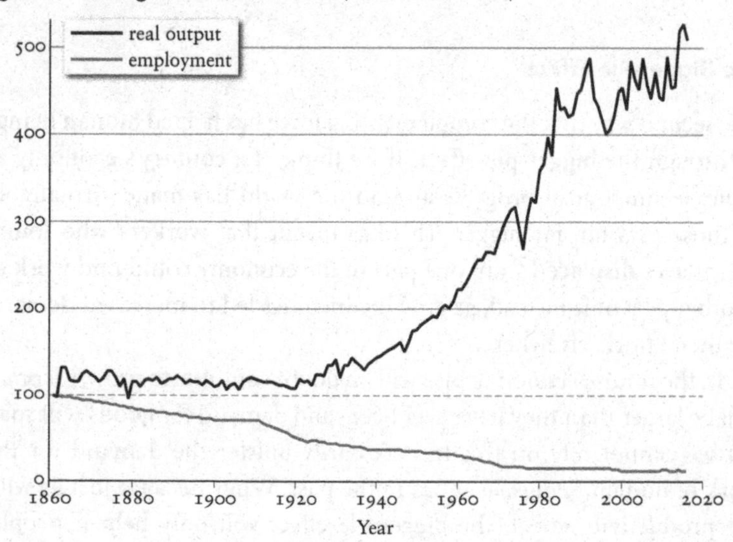

but as new technologies have spread, the demand for human beings to create it has shriveled.

Alternatively, take the UK manufacturing sector since 1948, shown in Figure 7.2. This part of the British economic pie also grew in the second half of the twentieth century. And while that initially led to more work for people to do, employment started to fall toward the end of the 1970s. Today, the sector produces about 150 percent more than it did in 1948, and yet requires 60 percent fewer workers to do it. Once again, the amount of money spent on manufacturing output is greater than it has ever been, but as new technologies have spread, demand for people to produce that output has fallen away.

None of this is unique to Britain, either. The same progression took place in US manufacturing, for example—a sector of the American economy that grew significantly over the last few decades but did not create more work for people to do. Today, it produces about 70 percent more output than it did back in 1986, but requires 30 percent fewer people to produce it. In the first decade of the twenty-first century alone, 5.7 million American manufacturing jobs were lost.[10]

Figure 7.2: UK Manufacturing, 1948–2016 (Index 1948 = 100)[11]

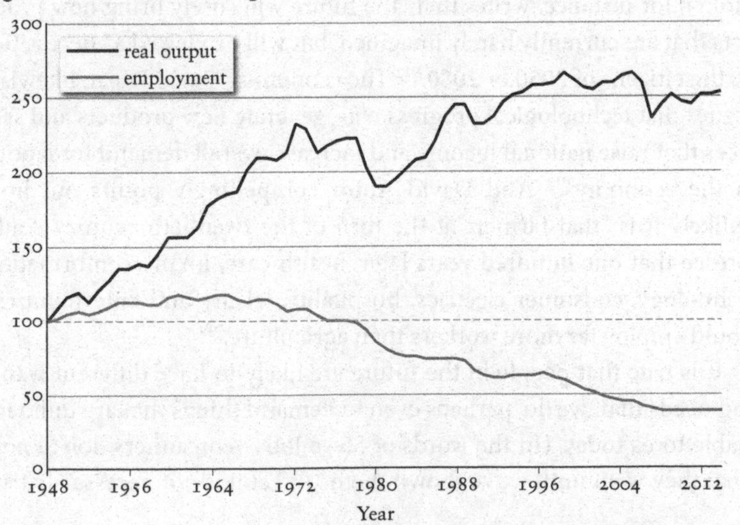

At the moment, these stories are only unfolding in particular parts of the economy. They are not, as yet, universal phenomena. But they still capture the essence of the problem with the bigger-pie effect: rising incomes may lead to rising demand for goods, but that does not necessarily mean rising demand for the work of human beings. In agriculture and manufacturing, in the UK and the United States, the two trends have already decoupled. The troubling thought is that, as task encroachment continues, this might happen in other areas of economic life as well.

The Changing-Pie Effect: Consumers

The final way that the complementing force has helped human beings in the past is through the changing-pie effect: technological progress not only made economic pies bigger, but it added entirely new ingredients to them. How did this happen? One way is that consumers not only had bigger incomes to spend, as before, but they changed how they spent those incomes as well. Over time, workers displaced from older jobs could find work producing these newly in-demand goods and services—including ones that had not even existed before.

When economists think about the future, the changing-pie effect is a particularly common source of optimism. The economic historian Joel Mokyr, for instance, writes that "the future will surely bring new products that are currently barely imagined, but will be viewed as necessities by the citizens of 2050 or 2080."[12] The economist David Dorn, likewise, argues that technological progress will "generate new products and services that raise national income and increase overall demand for labour in the economy."[13] And David Autor compellingly points out how unlikely it is "that farmers at the turn of the twentieth century could foresee that one hundred years later, health care, finance, information technology, consumer electrics, hospitality, leisure and entertainment would employ far more workers than agriculture."[14]

It is true that people in the future are likely to have different wants and needs than we do, perhaps even to demand things that are unimaginable to us today. (In the words of Steve Jobs, "consumers don't know what they want until we've shown them.")[15] Yet it is not necessarily true that this will lead to a greater demand for the work of human beings. Again, this will only be the case if human beings are better placed than machines to perform the tasks that have to be done to produce those goods. As task encroachment continues, though, it becomes more and more likely that changes in demand for goods will not turn out to be a boost in demand for the work of human beings, but of machines.

Looking at the newer parts of economic life, we might worry that something like this is already unfolding. In 1964, the most valuable company in the United States was AT&T, which had 758,611 employees. But in 2018 it was Apple, with only 132,000 employees; in 2019 it was overtaken by Microsoft, with 131,000. (Neither company existed in the 1960s.)[16] Or take a new industry like social media, populated by companies that are worth a great deal but employ comparatively few people. YouTube had only sixty-five employees when it was bought by Google for $1.65 billion in 2006; Instagram had just thirteen employees when it was bought by Facebook for $1 billion in 2012; and WhatsApp had fifty-five employees when it was bought by Facebook for $19 billion in 2014.[17] Research shows that in 2010, new industries that were created in the twenty-first century accounted for just 0.5 percent of all US employment.[18]

Such examples may turn out to be short-term empirical blips. Amazon, another one of today's most valuable companies, has about four and a half times as many employees as Apple or Microsoft (though still fewer than AT&T did at its peak). Nevertheless, those companies demonstrate that demand for goods in an economy can change dramatically, with entirely new industries emerging in response to meet those demands, and yet the demand for the work of human beings may still not rise. The disconcerting thought is that, once again, as task encroachment continues, this phenomenon will become more common.

The Changing-Pie Effect: Producers

There is also another way that the changing-pie effect has worked in the past: besides consumers buying different goods and services, producers have also changed the way that they make those goods and services

Figure 7.3: Horses and Mules and Tractors on US Farms, 1910–60 (000's)[19]

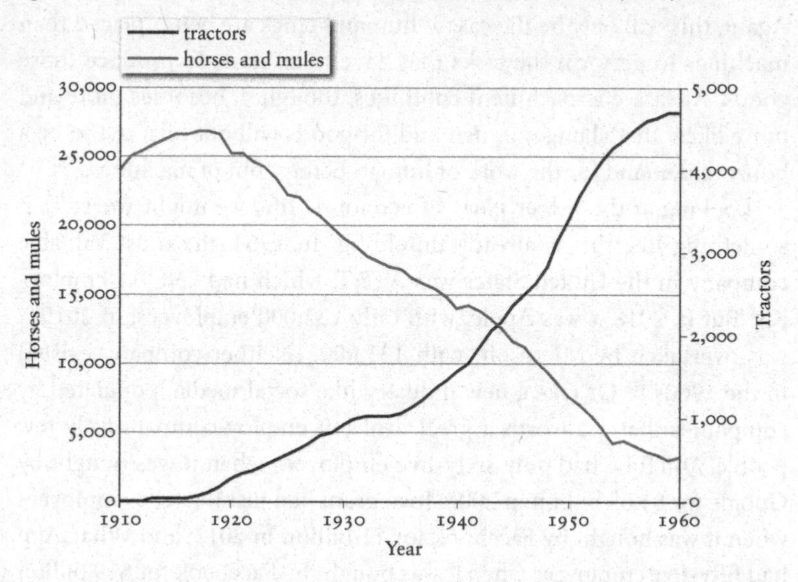

available. As old methods of production were overhauled by technological progress, people were displaced from their familiar tasks, but new tasks had to be done, and displaced workers were able to find jobs involving those activities instead.

For Daron Acemoglu and Pascual Restrepo, two leading economists, this version of the changing-pie effect provides a powerful response to Wassily Leontief's pessimism about the future of work.[20] His claim, remember, was that "what happened to horses will happen to people": just as tractors and cars put horses out of work, so new technologies would put human beings out of work as well.[21]

Acemoglu and Restrepo argue that there is an important difference between human beings and horses that explains why Leontief has been proven wrong up to now. As technological progress transformed the way that goods were produced, they suggest, this created a demand for "new and more complex tasks." Human beings, they say, are well suited to such activities, in a way that horses are not. And that is why they think Leontief was wrong to bundle people and horses together. As the economy changed, displaced human beings could change jobs and perform the complex new tasks that those jobs required. By contrast, displaced horses, fit only for pulling carts and carrying heavy loads, had nowhere else in the economy to go.[22]

For now, this assumption, that human beings are best equipped to perform new tasks that arise in the economy, may well still be right. But look further into the future, and it is far from clear that this will always be the case. Technological progress might indeed transform the way things are produced, requiring new tasks to be done—but why assume that these tasks will always be ones that human beings are best placed to do? As task encroachment continues, will it not become sensible to allocate more of the complex new tasks to machines instead?

Acemoglu and Restrepo provide one intriguing response.[23] They argue that when human beings are displaced by machines, human labor becomes cheaper: there are more workers looking for jobs, which pushes down their wages. That, in turn, creates an incentive for companies to invent new tasks for humans to do, to take advantage of these falling labor costs. And that is why human beings will in fact be best placed to perform new tasks: those tasks might be created precisely with them in mind.[24] It is an ingenious thought. But if it is right, it raises a new puzzle: Why did this mechanism not help out horses as well? When horses were displaced, they would have become cheaper, too. So why was there not a surge in the creation of new tasks that suited horses better than machines? Why did all these cheaper horses still lose their jobs?

The answer is that new tasks were not created for horses because their capabilities had been exhausted. No matter how cheap horses became, and how strong the incentive for entrepreneurs to take advantage of equine bargains, there was very little left for them to do that a machine could not do more efficiently. Horses became economically useless. This points to the problem with any argument that relies upon technological progress indefinitely creating new tasks for people to do in the future. At the moment, human capabilities are so impressive compared to those of machines that it is tempting to think that we can always find novel ways to put people to use. But as machines become more capable, in many areas of economic activity future human beings will look as feeble compared to machines as horses do today. More of the new tasks will be performed by machines instead. And the changing-pie effect will complement not us but them.

THE SUPERIORITY ASSUMPTION

There is a common thread running through all of the arguments above. Most of the time, when we think about the future of work, we imagine that human beings are special. We realize that as our economies grow and change, the demand for tasks to produce everything will grow and change as well. But all too often we simply take it for granted that people will remain the best choice to perform many of those tasks.

I call this the "superiority assumption." And when people appeal to the various aspects of the historically powerful complementing force as reasons to be optimistic about the future, you can see this assumption busily at work. We imagine that when *human beings* become more productive at a task, they will be better placed than a machine to perform it; that when the economic pie gets bigger, *human beings* will be better placed to perform the freshly in-demand tasks; that when the economic pie changes, *human beings* will be better placed to carry out whatever new tasks have to be done.

Until now, the superiority assumption has been a safe bet. If demand rose for some task, chances were that human beings were better placed than machines to do it. So that meant rising demand for workers, too. But as task encroachment continues, and machines take more tasks out the hands of human beings, that assumption looks increasingly dubious. Eventually, it will simply be wrong.

There is a line in the work of the nineteenth-century philosopher John Stuart Mill: "demand for commodities is not demand for labour."[25] Mill was not thinking about the future of work when he wrote this, but he may as well have been, because he is right: demand for "commodities," for goods and services, is not always demand for the work of human beings. In fact, it is only demand for whatever tasks have to be carried out to produce those commodities. If the superiority assumption holds, those tasks will indeed be done by human workers. But otherwise, they will not.

This means that the productivity effect may increase the demand for workers at a task—but once they are displaced by a more capable machine, this helpful effect withers away. The bigger-pie effect may increase the demand for workers at a task—but once they are displaced, this helpful effect disappears as well. And the changing-pie effect may increase the demand for workers at a task—but, again, once they are displaced by machines, this helpful effect fades, too.

WHAT REMAINS FOR US TO DO

But surely, one might object, there are some activities that machines will never do, even if they are extraordinarily capable. Couldn't everyone just find work doing those? Even in a world where human beings are left with only some residual tasks to do, might there not be enough demand for those few tasks to keep everyone employed?

It is indeed entirely plausible that some tasks will remain for us: those that prove impossible to automate, others that are possible but unprofitable to automate, and still others that are both possible and profitable to automate but remain restricted to human beings due to regulatory or cultural barriers that societies build around them. There are also tasks that might remain out of reach because we value the very fact that they are done by human beings, not a machine. That last reason explains why, in 2018, millions of people went online to watch Magnus Carlsen, the world chess champion, defend his title against Fabiano Caruana; machines could easily defeat either one of them, but spectators valued not only the particular movement of the chess pieces but the fact that they were being moved by human beings.[26] It also explains why diners in a fine restaurant might feel shortchanged if they discover that their coffee was made by a capsule-based machine rather than by a highly trained barista, even though capsule-based coffee is often preferred in blind tests: people value not only the taste, but the fact that a person brewed the coffee for them.[27] Throughout life, we can point to certain tasks—crafting furniture, tailoring a suit, preparing a meal, caring for one another in old age and ill health—where we value the process behind them, and in particular the fact that they are done by human beings, rather than just the outcome that is achieved.

And yet, though some such residual tasks might always remain for human beings to do, it is a mistake to think that there is likely to be enough demand for them to keep everyone in work. To picture the problem with that thought, imagine a large ball pit filled up with balls. Each ball represents a particular type of task in the economy, and every task has a ball in the pit to represent it. A ball is blue if a human being is best placed to take on the task in question, and red if the task is one more suitable for machines. Several hundred years ago, almost all of the balls in the ball pit would have been blue. As time has passed, though, more of the balls have become red. This is task encroachment at work.

Now imagine that each ball varies in size as well as color. Some are gigantic while others are tiny, their size reflecting the demand for that particular task in the economy. In principle, even in a world where only a few balls in the pit remain blue—that is, where human beings are best placed to perform only a handful of residual tasks—if those balls are large enough, then enough demand would remain for the work of human beings to keep everyone employed. For instance, if one of those tasks is handcrafting furniture, and there is immense demand for bespoke chairs and custom cabinets, it is possible that everyone could find work as a carpenter. It would be a very strange and highly monotonous world, but there would still be enough work for people to do.

But an economy populated entirely by craftspeople like this sounds absurd—and this fact is revealing. Even in the long run, when machines are unimaginably more capable than they are today, it would be unsurprising to learn that some tasks remain for human beings to do for all the reasons set out before. In other words, some blue balls will always remain in the pit. But it would be very surprising to find out that those balls are big enough to support everyone looking for work. It is possible that this could happen, but it seems quite unlikely—and as time passes and task encroachment continues, forcing human beings to retreat to fewer and fewer types of tasks, it becomes ever *more* unlikely. As more red spreads through the ball pit, it is increasingly improbable that the dwindling number of blue balls that remain turn out to be not just decent-sized, but so extraordinarily vast that they can provide employment for everyone.

Think again of British farmers. In a sense, they already find themselves in this position. Despite the technological progress that has taken place in agriculture over the last hundred-odd years, some tasks still remain for them to do today—but there is only enough demand for those tasks to keep about a tenth as many of the farmers employed today as there were in 1861. Or think of the British factory workers. Again, although production processes became increasingly automated in the second half of the twentieth century, some tasks still remain for human beings to do—yet again, there is only enough demand now for those tasks to keep about 40 percent as many factory workers employed as there were in 1948.

In a similar way, we might imagine, for very good reasons, that many other jobs that exist today—and those that do not yet exist as well—will turn out in the future to involve some tasks that are best done by humans rather than machines. But as those tasks dwindle in number, there is no reason to think there will be enough demand for them to keep everyone at work in those roles.

"THE LUMP OF LABOR FALLACY" FALLACY

The arguments above also pose a problem for the so-called lump of labor fallacy—a fallacy that people are often accused of committing when they seem to forget about the helpful side of technological progress, the complementing force.[28] The idea is an old one, first identified back in 1892 by David Schloss, a British economist.[29] Schloss was taken aback when he came across a worker who had begun to use a machine to make washers, the small metal discs used when tightening screws, and who appeared to feel *guilty* about being more productive. When asked why he felt that way, the worker replied: "I know I am doing wrong. I am taking away the work of another man."

Schloss came to see this as a typical attitude among workmen of the time. It was, he wrote, a belief "firmly entertained by a large section of our working-classes, that for a man . . . to do his level best—is inconsistent . . . with loyalty to the cause of labour." He called this the "theory of the Lump of Labour": it held "that there is a certain fixed amount of work to be done, and that it is best, in the interests of the workmen, that each man shall take care not to do too much work, in order that thus the Lump of Labour may be spread out thin over the whole body of workpeople."[30]

Schloss called this way of thinking "a noteworthy fallacy." The error with it, he pointed out, is that the "lump of work" is in fact not fixed. As the worker became more productive, and the price of the washers made by him fell, demand for them would increase. The lump of work to be divided up would get bigger, and there would actually be more for his colleagues to do.

Today, this fallacy is cited in discussions about all types of work. In its most general terms, it is used to argue that there is no fixed lump of work in the economy to be divided up between people and machines; instead, technological progress raises the demand for work performed by everyone in the economy. In other words, it is a version of the point that economists make about the two fundamental forces of technological progress: machines may substitute for workers, leaving less of the original "lump of work" for human beings, but they complement workers as well, increasing the size of the "lump of work" in the economy overall.

As this chapter shows, however, there is a serious problem with this argument: over time, it is likely to become a fallacy itself. (We might call this the "'Lump of Labor Fallacy' Fallacy," or "LOLFF" for short.) It may be right that technological progress increases the overall demand for work. But it is wrong to think that human beings will necessarily be better placed to perform the tasks that are involved in meeting that demand. The lump of labor fallacy involves mistakenly assuming that the lump of work is fixed. But the LOLFF involves mistakenly assuming that that growth in the lump of work has to involve tasks that human beings—not machines—are best placed to perform.

A WORLD WITH LESS WORK

We can now begin to see how the Age of Labor is likely to end. As time goes on, machines continue to become more capable, taking on tasks that once fell to human beings. The harmful substituting force displaces workers in the familiar way. For a time, the helpful complementing force continues to raise the demand for those displaced workers elsewhere. But as task encroachment goes on, and more and more tasks fall to machines, that helpful force is weakened as well. Human beings find themselves complemented in an ever-shrinking set of tasks. And there is no reason to think the demand for those particular tasks will be large enough to keep everyone employed. The world of work comes to an end not with a bang, but a withering—a withering in the demand for the work of human beings, as the substituting force gradually overruns the complementing force and the balance between the two no longer tips in favor of human beings.

There is no reason, though, to think that the demand for the work of human beings will dry up at a steady pace. There may be sudden surges in either force: a burst of worker displacement here, a surge in demand for workers there. Nor will the demand for the work of human beings dry up at the same pace in all parts of an economy. Some industries might be more exposed to one force than the other; some regions will be more insulated than others. And it is also important to remember that any fall in demand for the work of human beings may initially not change the amount of work to be done, but the nature of that work: its pay, its quality, its status. In the end, though, it is the number of jobs that will be affected. As Leontief put it, lowering workers' wages could "postpone [their] replacement by machines for the same reason that a reduction of oats rations allocated to horses could delay their replacement by tractors. But this would be only a temporary slowdown in the process."[31]

As machines keep becoming increasingly capable, many human beings will eventually be driven out of work. In fact, some economists have already seen this happening in the data. When Daron Acemoglu and Pascual Restrepo looked at the use of industrial robots in the United States from 1990 to 2007, they found a contemporary case of the substituting force overrunning the complementing force, reducing the demand for workers across the entire economy. Remember that in thinking about new technologies, we are used to stories like that of the ATM: machines displace some people, but also raise the demand for their work elsewhere, and overall employment stays the same or even rises. But that is not what happened with industrial robots. On average, one more robot per thousand workers meant about 5.6 fewer jobs in the entire economy, and wages that were about 0.5 percent lower across the whole economy as well. And all this was happening in 2007, more than a decade ago, before most of the technological advances described in the preceding pages.[32]

Critics might point out that this result applied not to all technologies, but only to one particular category, industrial robots. But that misses the deeper point: traditionally, many economists have imagined that this result was not possible for *any* technology. The illusion nurtured in the Age of Labor was that any technological progress ultimately benefits workers overall. But here, even after taking into account the ways in which these industrial robots helped some human beings through the complementing force, workers overall were still worse off.

THE TIMING

How long will it take to arrive at a world with less work? It is very hard to say precisely. That is not meant to be evasive: I truly do not know the answer. Our pace of travel will depend upon the accumulated actions of an unimaginably large number of individuals and institutions, each with their own part to play on the economic stage—the inventors who create the technologies and the companies that put them to use, the workers who decide how to interact with them and the state that figures out how to respond, to name but a few. All we know with any degree of certainty is that tomorrow's machines will be more capable than they are today, taking on more and more tasks once performed by human beings. Statements like "X percent of people will be unemployed Y years from now" may be reassuringly unambiguous, but such straightforward predictions, however sophisticated the reasoning behind them, are likely to be a misleading guide to the future of work.

Nevertheless, there are some general observations that can be made about timings. Roy Amara, an influential Silicon Valley figure, once said that "we tend to overestimate the effect of a technology in the short run and underestimate the effect in the long run."[33] This is a helpful way to think about what lies ahead. Current fears about an imminent collapse in the demand for the work of human beings are overblown. In the short run, our challenge will be avoiding frictional technological unemployment: in all likelihood, there will be enough work for human beings to do for a while yet, and the main risk is that some people will not be able to take it up. But in the longer run, in the spirit of Amara's reflection, we have to take seriously the threat of structural technological unemployment, where there is simply not enough demand for the work of human beings.

But how distant is that threat? Why bother worrying about it if, as Keynes famously quipped, "in the long run we are all dead"? In writing about technology and the long run, I have in mind decades, not centuries. On that count, I am more optimistic than Keynes: I hope that my readers and I will be alive to see the long run unfurl. And even if we are not, our children certainly will be. For their sakes, at the very least, we need to take the problem of a world with less work very seriously. Frictional technological unemployment is already becoming evident, and in some corners of economic life today we can catch glimpses of how structural technological unemployment might emerge. Given technological trends, it is hard to imagine that these challenges will not become more and more acute as time goes on. It is deeply sobering to think, for instance, that if technological progress continues at the same rate for the next eight decades as over the last eight, then our systems and machines—which have already accomplished so much, and surprised so many—will be a further trillion times more powerful by 2100 than they already are today. To be sure, no trend is certain to last, and computational power is not everything. But progress that even faintly resembles this will, I believe, make technological unemployment a critical challenge within the twenty-first century.

What's more, we do not have to wait for vast pools of human beings to be displaced before technological unemployment becomes a problem. Much of the current conversation about the future of work assumes that we only need worry when *most* people are left without work. But even in a world where just a sizable minority of human beings—perhaps 15 or 20 percent—find themselves in that position, we should already be concerned about the instability that inaction might bring. Remember that in 1932, a rise in the German unemployment rate to 24 percent helped bring Hitler to power.[34] This was not, of course, the only reason for his success: other countries, with similar labor market stories, did not turn to fascism in response. But the German experience should make us all sit up a little straighter.

ALL ABOUT US

Today some may laugh at Leontief, who warned that human beings would face the same fate as horses—unemployment. But in the decades to come, I imagine, he will be laughing at us, from the academy of economists in the sky. Like Keynes and his predictions about technological unemployment, Leontief may have misjudged the timing but, with great foresight, he recognized the final destination. Just as today, we talk about "horsepower," harking back to a time when the pulling power of a draft horse was a measure that mattered, future generations may come to use the term "manpower" as a similar kind of throwback, a relic of a time when human beings considered themselves so economically important that they crowned themselves as a unit of measurement.

I have spoken about our "superiority assumption" when it comes to comparing human capabilities with those of machines. But eventually an *inferiority assumption* will be a better starting point for thinking about technology and work—where machines, not human beings, become the default choice to perform most activities. Economists have built up an impressive arsenal of reasoning to explain why there will always be enough demand for the work of human beings. But as we have seen, all those arguments depend upon human beings remaining best placed to perform whatever tasks are in demand as our economies grow and change. Once that comes to an end, and machines take our place, all that argumentative firepower will turn back on us, explaining instead why there will always be a healthy demand for the work of machines rather than human beings.

The threat of technological unemployment set out in this chapter might sound extraordinary, in the literal sense of the word: an out-of-the-ordinary phenomenon that is entirely unrelated to life today. But as we will see in a moment, this is not quite right. Instead, the threat is best thought of as a more extreme version of something that is already affecting us right now: the problem of rising inequality.

8

Technology and Inequality

Economic inequality is a phenomenon as old as civilization itself. Prosperity has always been unevenly shared out in society, and human beings have always struggled to agree on what to do about that.

It is tempting to imagine that this is not so. The eighteenth-century philosopher Jean-Jacques Rousseau, for example, believed that if you went back far enough, you would find human beings leading a "simple and solitary life," free from any "chains of dependence" on one another. In his writings on "The Genesis of Inequality," he pictures himself back in this "state of nature," unfettered by anyone else's demands. If a fellow human being tries to impose any work on him, Rousseau says, "I take twenty steps into the forest; my chains are broken, and he never sees me again for the rest of his life."[1] Once upon a time, in this vision, people could dodge the challenge of inequality simply by turning away and retreating into solitude.

But this philosophical fiction is misleading. In fact, from what we know about some of our earliest ancestors, the hunter-gatherers who roamed the African savannah hundreds of thousands of years ago, such a retreat was never possible.[2] It is true that the hunter-gatherers did not live in large, stable societies like our own. Their economic pies were smaller, if it makes sense to think of there being "economies" back then at all.

And material inequalities between them were narrower: major divides would only emerge after the last Ice Age, about twelve thousand years ago, when the climate became more stable, farming and herding spread, and some people were able to build up resources that others did not.[3] Yet even so, the hunter-gatherers did not pursue solitary lives of the kind that Rousseau imagined. Instead, they lived together in tribes that sometimes numbered a few hundred people, sharing the literal fruits (and meats) of their labor within their band of fellow foragers—some of whom, inevitably, were more successful in their foraging efforts than others.[4] There is no forest that lets human beings retreat into perfect solitude and self-sufficiency, nor has there ever been. All human societies, small and large, simple and complex, poor and affluent, have had to figure out how best to share their unevenly allocated prosperity with one another.

Over the last few centuries, humanity's collective prosperity has skyrocketed, as technological progress has made us far wealthier than ever before. To share out those riches, almost all societies have settled upon the market mechanism, rewarding people in various ways for the work that they do and the things that they own. But rising inequality, itself often driven by technology, has started to put that mechanism under strain. Today, markets already provide immense rewards to some people but leave many others with very little. And now, technological unemployment threatens to become a more radical version of the same story, taking place in the particular market we rely upon the most: the labor market. As that market begins to break down, more and more people will be in danger of not receiving a share of society's prosperity at all.

THE TWO TYPES OF CAPITAL

People who are frustrated with the current economic system sometimes say that "the problem with capitalism is that not everyone has capital." The complaint, in other words, is that income today only flows reliably to those who own "things," like stocks and shares, real estate and patents. Leaving aside for the moment the question of whether that accusation is correct, the statement contains an important, revealing mistake: the idea that the only type of capital in the world involves ownership of some kind of property, what economists might call *traditional* capital.

In the words of economist Thomas Piketty, traditional capital is "everything owned by the residents and governments of a given country at a given point in time, provided that it can be traded on some market." It is a broad definition that captures things tangible and intangible, financial and nonfinancial: land, buildings, machinery, merchandise, intellectual property, bank accounts, equities and bonds, software and data, and so on.[5] The complaint from above is right to point out that not all people own such traditional capital, but it is wrong to conclude that they therefore own no capital at all. Everyone in the world does own another type of capital: themselves.

Economists call this *human* capital, a term meant to capture the entire bundle of skills and talents that people build up over their lives and put to use in their work. Arthur Pigou was the first economist to use the phrase, back in the 1920s; decades later, Gary Becker won a Nobel Prize for his work on the topic.[6] Its name follows from the family resemblances it has with traditional capital: people can invest in it (through education), some kinds of it are more valuable than others (e.g., specialized skills), and when put to use it provides a return to its owner (in the form of a wage). Unlike traditional capital, though, human capital is stored up inside us and cannot be traded on a market—unless, of course, its owner comes along with it.

To some the idea of human capital might sound overly mechanistic, an economic abstraction far removed from real life. In his Nobel Prize lecture, Becker describes how he ran into this reaction when he began his work: "the very concept of *human* capital was alleged to be demeaning because it treated people as machines. To approach schooling as an investment rather than a cultural experience was considered unfeeling and extremely narrow."[7] But by demystifying human beings in this way, by not granting them magical powers that make them unlike anything else in the economy, the idea of human capital helps us to think clearly about the challenge that lies ahead.

THE CHALLENGE OF TECHNOLOGICAL UNEMPLOYMENT

Looked at in this light, technological unemployment is what happens when some people find themselves with human capital that is of no value in the labor market—that is, when no one wants to pay them to put their skills and talents to use. This is not to say they will not *have* any human capital. They almost certainly will, having spent large parts of their lives acquiring education or training, perhaps at great effort and expense. The problem is that, in a world with less work, this human capital might turn out to be worthless. In the case of frictional technological unemployment, the human capital may be of the wrong type for the available work; in the case of structural technological unemployment, there may not be enough demand for human capital at all.

What's more, as we have seen, there are two types of capital: people earn money both as a return on the human capital they have built up, and also on any traditional capital that they hold. In a world with less work, the flow of income many people receive from their work may dry up to a trickle, but the flow of income going to those who own the latest systems and machines—the new forms of traditional capital that displaced those workers in the first place—is likely to be quite considerable.

Now, if everyone happened to own a portfolio of traditional capital along those lines, we would probably worry far less about the prospect of a world with less work. After all, that is the story of the British aristocracy over the last few centuries. As George Orwell warmly put it, they are "an entirely functionless class, living on money that was invested they hardly knew where . . . simply parasites, less useful to society than his fleas are to a dog."[8] Economically speaking, they are a relatively useless group of people, yet because of their handsome holdings of traditional capital they still receive a good deal of income.

It is unlikely, though, that those who find themselves out of work in the future will be able to follow their example. For most, the prospect is not the prosperous life of a propertied aristocrat, but an existence with little to no income at all. A world with less work, then, will be a deeply divided one: some people will own vast amounts of valuable traditional capital, but others will find themselves with virtually no capital of either kind.

A world like this is not science fiction. In fact, it looks a lot like a more extreme account of the one we live in today: income already flows to different people at very different rates, pouring into the pockets of some but barely trickling to others. And this resemblance is not a coincidence. The phenomena of inequality and technological unemployment are very closely related. Most societies have decided to slice up their economic pies by using the market to reward people for whatever capital they happen to own, whether human or traditional. Inequality is what happens when some people have capital that is far less valuable than that of others; technological unemployment is what happens when some have no capital worth anything in the market at all—certainly no human capital of worth, and likely no traditional capital, either.

Exploring existing inequalities, then, is useful because it shows us how a world with insufficient work can emerge from what we already see around us. In a sense, today's inequalities are the birth pangs of tomorrow's technological unemployment.

INCOME INEQUALITY

How can we see the current trends in inequality? One way is to look at inequality of overall incomes, and in particular something known as the Gini coefficient. This is a number that captures how incomes are spread out: if everyone has the same income in a particular society, then its Gini coefficient is zero, and if only one person earns everything, then the Gini coefficient is one.[9] In most developed countries, this number has risen significantly over the last few decades.[10] (In less-developed countries the story is somewhat ambiguous: their Gini coefficients were usually very high to begin with, but have remained relatively stable.) In other words, the largest economic pies, belonging to the most prosperous nations, are being shared out less equally than in the past.

There is some disagreement, though, about the usefulness of the Gini coefficient. Boiling everything down to a single number has an attractive simplicity to it, but inevitably important details are lost along the way.[11] A different approach might be to look at the full spread of incomes in a particular economy. By lining up all the different income groups from lowest to highest, and noting how each group's incomes have changed over time, it is possible to get a sense of how things are changing across the entire distribution. If you look at the United States, for example, the results are striking.

Figure 8.1: Gini Coefficients from Mid-1980s to 2017 (or Latest)[12]

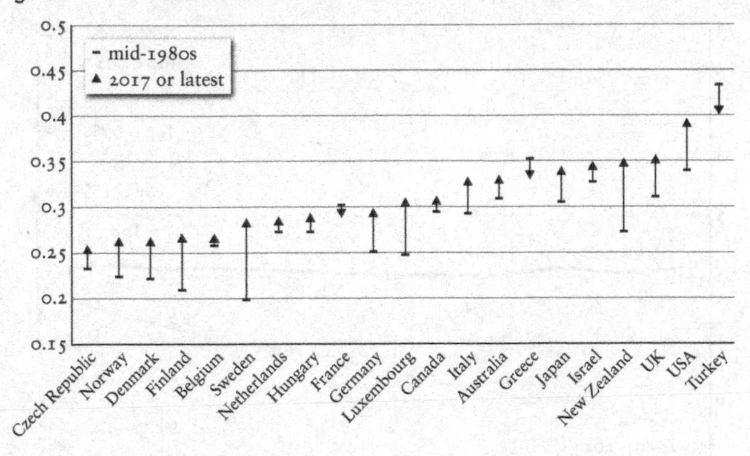

Figure 8.2 shows that in the thirty-four years before 1980, income growth in the United States was fairly solid for everyone. In the thirty-four years after 1980, however, income growth was anemic for the people who earned the least, while it soared for the 1 percent who earned the most. For followers of John Rawls, the vastly influential twentieth-century political philosopher, this picture is especially off-putting. In his great work, *A Theory of Justice*, Rawls argued that inequalities should "be to the greatest benefit of the least advantaged" members of society.[13] Before 1980, the spirit of that principle was largely upheld in this economic domain: the incomes of the poorest increased as much as those of others, or even slightly more. Today, though, as Figure 8.2 shows, the reverse is true: it is the incomes of the richest that have risen instead.

Focusing our attention on that richest fraction of society gives us a third approach to the issue, known as "top income inequality" or just "top inequality." This measure has captured the imagination of protesters and public commentators in the last decade, with "The One Percent" becoming a well-known label and "We are the 99 Percent" the battle cry of the Occupy Movement. Their frustration is not without cause: the proportion of total income that goes to the 1 percent who earn the most, particularly in developed countries, has increased significantly. In the United States and the UK, that share has almost doubled over the last

Figure 8.2: Average Annual US Income Growth[14]

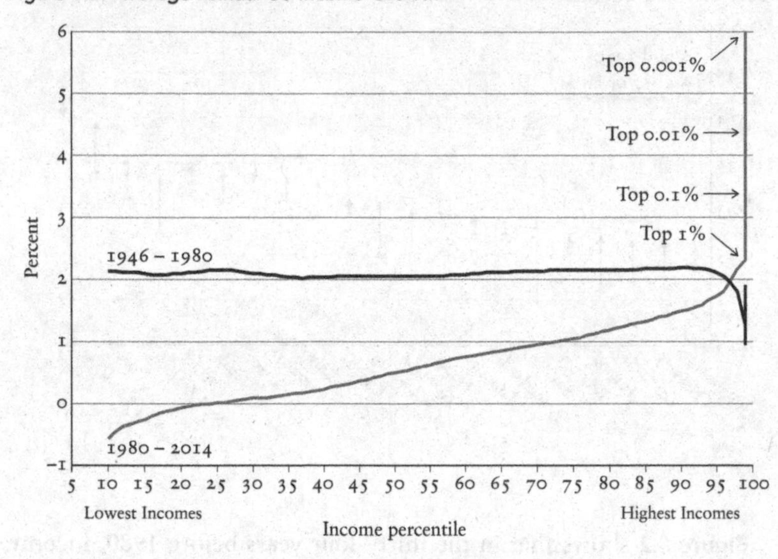

few decades.[15] Figure 8.3 shows that much the same story is unfolding elsewhere.

Even in Nordic countries, like Finland, Norway, and Sweden, often lauded for their equality, the share going to the richest 1 percent has grown. And if we narrow our focus further, to look just at the top 0.1 and 0.01 percent, the picture is often more extreme. In the United States, for example, from 1981 to 2017, the income share of the top 0.1 percent increased more than three and a half times from its already disproportionately high level, and the share of the top 0.01 percent rose more than fivefold.[16]

The three measures of income inequality do diverge sometimes, of course, and it is possible to point to particular instances where some measurement departs from an upward trend. The Gini coefficient in the UK, for instance, has not really risen for twenty-five years.[17] But it is rare to find a country where none of the three measurements show rising inequality; in the UK, as Figure 8.3 shows, the income share of the 1 percent has soared. And when all the measures, applied to lots of different countries, are taken together, the big picture is clear: in the most

Figure 8.3: Income Shares of the Top 1 Percent from 1981 to 2016 (or Latest)[18]

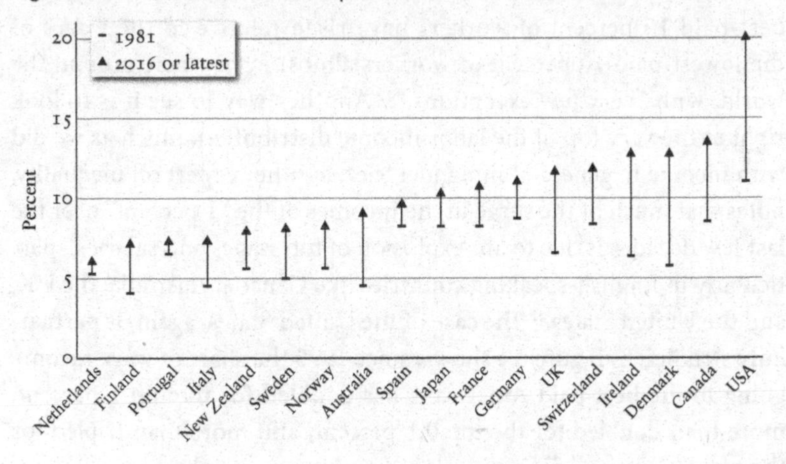

prosperous parts of the world, we are seeing a move toward societies with greater income inequality.

Why is income inequality rising, though? The short answer is that valuable capital is being shared out in an increasingly unequal way. As a result, the income that flows to those who hold that capital is increasingly unequal, too. More specifically, rising income inequality comes from increasingly unequal returns both on human capital and on traditional capital. Let us take each of those in turn.

INEQUALITY IN LABOR INCOME

Many people own little besides their human capital, the various skills they have acquired during their lives. Their paychecks are therefore their main source of income. Indeed, in many countries salaries and wages make up about three-quarters of the total income in the economy.[19] It should therefore come as no surprise that much of the general rise in income inequality from before is rooted in a rise in *labor income* inequality in particular. In other words, inequality is rising because workers are being paid more and more unequally for their efforts.[20]

One way to see that labor income inequality is rising is to compare different income deciles. Anthony Atkinson, a leading scholar of inequality, found that over the last few decades, the wages of the best-paid 10 percent of workers have risen relative to the wages of the lowest-paid 10 percent of workers almost everywhere around the world, with "very few exceptions."[21] Another way to see it is to look right at the very top of the labor income distribution, much as we did with income in general. Emmanuel Saez, another expert on inequality, notes that much of the surge in the incomes of the "1 percent" over the last few decades is due to an "explosion of top wages and salaries," particularly in English-speaking countries like Canada, Australia, the UK, and the United States.[22] The case of the United States, again, is particularly striking: as Figure 8.4 shows, since 1970 the share of wage income going to the best-paid Americans has doubled for the top 1 percent, more than doubled for the top 0.1 percent, and more than tripled for the top 0.01 percent.[23]

Notably, much of this rising inequality in labor income is due specifically to technological progress. As we saw in chapter 2, starting in the second half of the twentieth century new technologies were responsible for widening the gap between the wages of well-educated workers and everyone else, an increase in the so-called skill premium. This is an important part of the explanation for why the top 10 percent of wage earners have done so well in so many countries.

Right at the very top, though, the role of technological progress in causing labor income inequality is less clear. Some economists do believe that new technologies are directly responsible for the rising pay of the top 1 percent and 0.1 percent of wage earners. For instance, CEOs are thought to use new systems to run larger, more valuable companies, pushing up their pay as a result. Bankers, who stand alongside CEOs at the top of the pay ladder, may also have seen technological progress boost their wages, as financial innovations like complex pricing software and algorithmic trading platforms have helped to raise the demand for their work.[24]

The most compelling explanations for the rise in inequality at the very top, though, are not so much about productivity but *power*: these "supermanagers," as Thomas Piketty calls them, are receiving higher wages largely because they now have so much institutional clout that they are able to put together increasingly generous pay packages for

Figure 8.4: Growth in the Top US Wage Income Shares (Index 1970 Share = 100)[25]

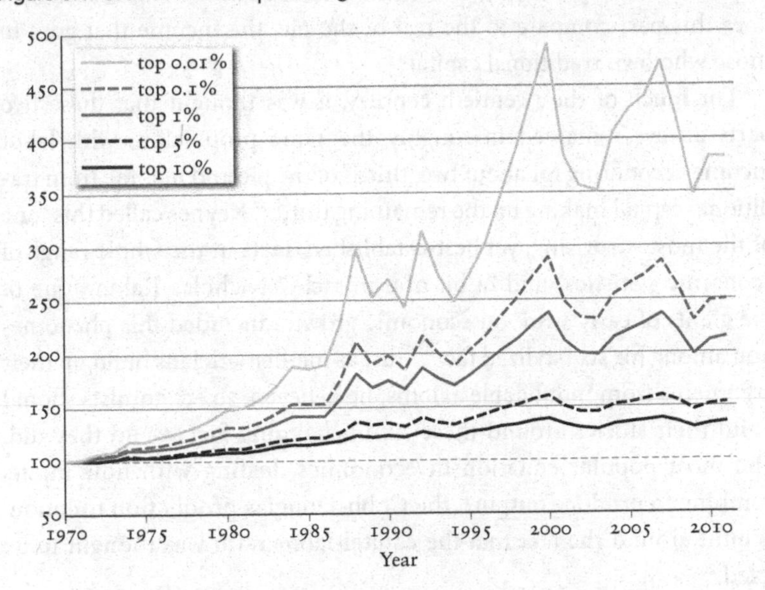

themselves. In this account, technological progress does have a role in making the economic pie bigger, but the growing power of these super-managers also allows them to take a much bigger slice of it. Forty years ago, the CEOs of America's largest firms earned about 28 times more than an average worker; by 2000, that ratio stood at an astounding 376 times.[26] At that point, a top CEO earned more in a day than an average worker did in an entire year.

Such wage inequality can be shocking, but there is also an optimistic reading of this trend: it shows that the lopsidedness of labor income is not inevitable. If powerful people can influence their pay in this way, then we do not have to treat economic imbalances as something beyond human control. In this case, power is being used to increase inequality, but using it to do the reverse is possible, too. We will return to this thought at the end of the chapter.

INEQUALITY BETWEEN LABOR AND CAPITAL

The part of the economic pie, then, that goes to workers in the form of wages and salaries is being sliced up in an increasingly unequal way:

some get a far larger return on their human capital than others. But how does this part compare to the rest of the pie, the income that goes to those who own traditional capital?

For much of the twentieth century, it was thought that these two parts always remained in roughly the same proportion, with labor income accounting for about two-thirds of the pie and income from traditional capital making up the remaining third.[27] Keynes called this "one of the most surprising, yet best-established, facts in the whole range of economic statistics" and "a bit of a miracle." Nicholas Kaldor, one of the giants of early work on economic growth, included this phenomenon among his six "stylized facts." Just as mathematicians build up their arguments from indubitable axioms, he believed, so economists should build their stories around these six unchanging facts—and they did. The most popular equation in economics dealing with how inputs combine to produce outputs, the Cobb-Douglas production function, is built around the fact that the capital-labor ratio was thought to be fixed.[28]

Until recently, Keynes's "miracle" has held true. But around the world, in the last few decades, the part of the pie that goes to workers (what economists call the "labor share") has begun to shrink, and the part that goes to those who own traditional capital (the "capital share") has started to grow.[29] In developed countries, this trend has been under way since the 1980s, and in developing countries, since the 1990s.[30]

Why has the labor share fallen like this? As productivity has risen around the world, and economic pies have gotten bigger, an ever diminishing part of that growth in income has flowed to workers through their wages. In the two decades since 1995, across twenty-four countries, productivity rose on average by 30 percent, but pay by only 16 percent.[31] Instead of going to workers, the extra income has increasingly gone to owners of traditional capital. This "decoupling" of productivity and pay, as it sometimes known, is particularly clear in the United States, as seen in Figure 8.6. Until the early 1970s, productivity and pay in the United States were almost perfect twins, growing at a similar rate. But as time went on, the former continued upward while the latter stalled, causing them to diverge.

A great deal of this decline in the labor share, once again, is due to

Figure 8.5: Falling Labor Share of Income in Developed Economies[32]

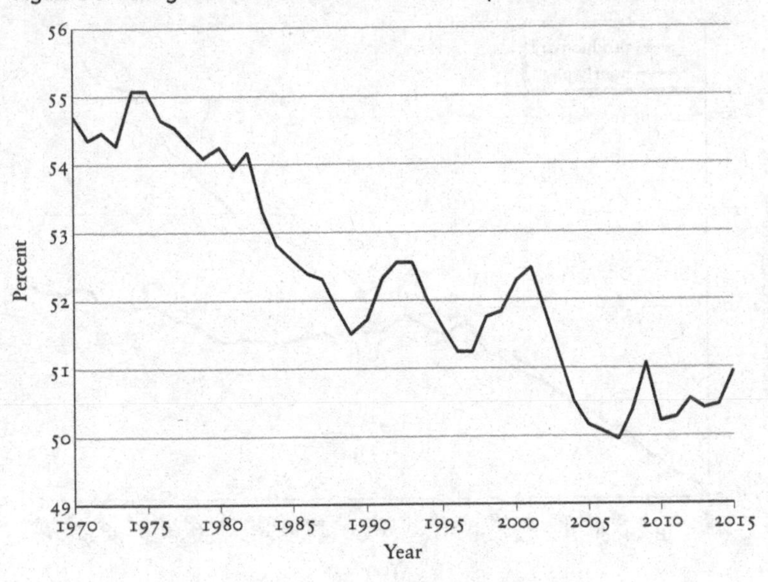

technological progress. The OECD is quoted as saying that technology was directly responsible for up to 80 percent of the decline from 1990 to 2007, encouraging firms to shift toward using more traditional capital relative to labor.[33] The IMF puts it at a more modest 50 percent in developed economies over a slightly longer period, a finding that fits with the work of other economists.[34] But once you look at the explanations offered by the IMF for the rest of the decline, technological progress often has a role to play there as well. Part of this decline in the labor share, for instance, is thought to be explained by globalization, the increasingly free movement of goods, services, and capital around the world. The IMF believes that this explains another 25 percent.[35] But what is actually responsible for this globalization? Technological progress, in large part. After all, it is falling transportation and communication costs that have made globalization possible.

Another explanation for the fall in the labor share is the rise of "superstar" firms: a small number of highly profitable companies that are more productive than their competitors and capture a large share of their respective markets. These superstar firms tend to require less

Figure 8.6: US Productivity and Wages, 1948–2016[36]

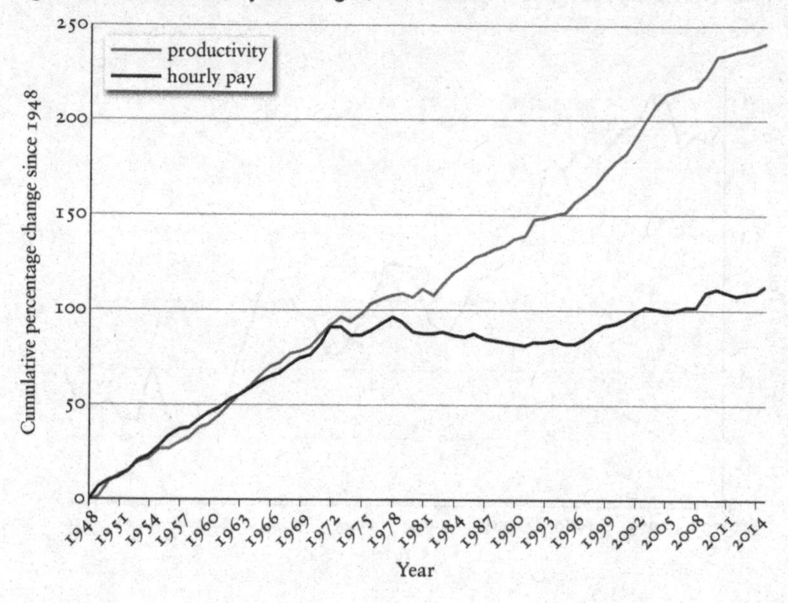

labor input per unit of output than their displaced competitors, and so as they become more dominant the labor share of income in the overall economy falls. Sales concentrations have risen, for instance, within most industries in the US private sector: and in the industries where that rise is the largest, the labor share has fallen the most.[37] Once again, though, when you look at the reasons for their dominance, technology—in the form of globalization, among other factors—is often thought to be responsible.[38] It is also telling that in industries where technological progress is faster, the concentration of companies appears to be rising most rapidly.[39] In 2018, of the top ten most valuable companies in the world, seven were in the technology sector.[40]

INEQUALITY IN CAPITAL INCOME

As the part of the pie that goes to workers shrinks, the part that goes to owners of traditional capital gets correspondingly larger. If we are concerned about inequality, this is a particularly problematic trend because the income from traditional capital is even more unevenly shared out across society than the income from salaries and wages. This fact is true "without exception," notes Thomas Piketty, in all countries and at all times for which data is available.[41]

The reason why the flow of income from traditional capital is so unequal is because the ownership of that capital is itself very unequally distributed—and increasingly so. In 2017, the charity Oxfam asserted that the world's eight richest men appeared to have as much wealth as the entire poorest half of the global population.[42] It is possible to argue about the details of its calculations, but other numbers paint a similar picture.[43] Gini coefficients for wealth, for instance, tend to be twice as high as those we saw calculated for income before.[44] And Piketty notes that in most countries, the richest 10 percent tend to own half or more (often much more) of all the wealth, while the poorest half of the country's population "own virtually nothing."[45]

The United States, once again, offers a particularly clear example. The poorest 50 percent of Americans own only 2 percent of the country's wealth.[46] But the richest 1 percent, who owned a bit less than 25 percent of the country's wealth at the end of the 1970s, now own more than 40 percent of it.[47] And the richest 0.1 percent of Americans, a group of just about 160,000 people, own about 22 percent of all the wealth in the country, with more than half of all wealth built up in the United States from 1986 to 2012 flowing to them.[48] The result is seen in Figure 8.7: America now finds itself in a situation where the top 0.1 percent holds the same amount of wealth as the poorest 90 percent combined. It is a throwback to an old-fashioned, 1930s arrangement of society, where a capital-holding class lives a gilded lifestyle while most people have very little by comparison.[49]

LOOKING AHEAD

As the preceding pages show, beneath the headline story of growing inequality around the world lie three distinct trends. First, human capital is less and less evenly distributed, with people's different skills getting rewarded to very different degrees; the part of the economic pie that goes to workers as a wage is being served out in an increasingly imbalanced

Figure 8.7: Wealth Shares of the Top 0.1 Percent and Bottom 90 Percent[50]

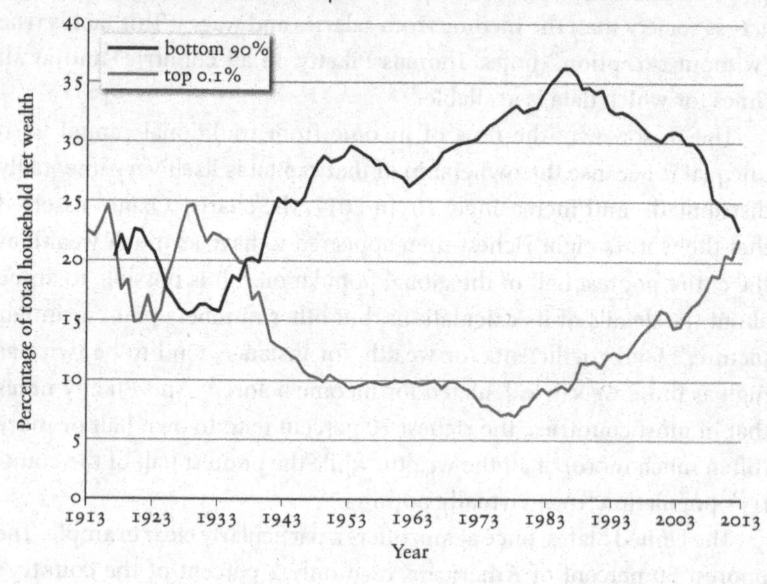

way. Second, human capital is becoming less and less valuable relative to traditional capital; that part of the pie that goes to workers as a wage is also shrinking relative to the part that goes to owners of traditional capital. And third, traditional capital itself is distributed in an extraordinarily uneven fashion, an inequality that has been growing more and more pronounced in recent decades.

These trends do not always hold everywhere in exactly the same way. In the UK, for instance, it is not clear whether the labor share of income has in fact fallen. In Central Europe and Japan, unlike the United States, the share of income going to the top 1 percent has actually tended to decrease during the twentieth century. There are other noteworthy exceptions, too. Still, the big picture remains the same: for the most part, economies around the world are becoming more prosperous but also more unequal. And the main culprit in this is technological progress.

Of course, technology is not the cause of all inequality. But it is often the main driver of these trends, both directly—by increasing the pay of high-skilled workers, or encouraging firms to use more traditional capital relative to labor—and indirectly, by promoting globalization and other economic shifts. (It is also worth remembering that technological progress is largely responsible for making the economic pie bigger in the first place. In other words, to a significant extent it accounts for both the "income" *and* the "inequality" in income inequality.)

Exploring how inequality works today is useful because it should help alleviate any residual skepticism about the looming threat of technological unemployment in the future. Right now, most societies share out prosperity by rewarding people in the marketplace for the capital that they own, both human and traditional. Today's growing inequalities show that this approach is already creaking: a few people own immensely valuable capital, but many more have little of value. Technological unemployment, as noted before, is just a more radical version of this same story; one where the market mechanism fails completely, and many people are left without any valuable capital at all. President John F. Kennedy famously quipped that a "rising tide lifts all boats," by which he meant that economic growth would benefit everyone in society. What he failed to point out, though, was that in a strong enough tide, those who find themselves without a boat—that is, those without any capital at all that is valued by the market—will simply drown.

But investigating how inequality works today is also useful for a second, more optimistic reason: it shows us that we can do something about technological unemployment in the future. As noted above, the three trends that underlie increasing inequality do not show up in the same way in all countries, even though everywhere is exposed to the same technological changes. This is not an inconvenient fact to be brushed aside, but a revealing one, suggesting that what countries do in response to technological change really matters. As the leading scholars of inequality put it, "income inequality has increased in nearly all world regions in recent decades, but at different speeds. The fact that inequality levels are so different among countries, even when countries share similar levels of development, highlights the important roles that national policies and institutions play in shaping inequality."[51]

Income inequalities are not inevitable. The only thing that is inevitable is that when some people arrive in this world, the lottery of life might or might not have granted them some unique talents and abilities, might or might not have placed them with particularly pleasant and affluent parents. These sorts of initial imbalances are unavoidable. But that is not true of all the imbalances that follow. There is no reason why those who are born with good luck should be the only people able to gather up valuable capital, nor any reason why the children of parents who enjoyed good fortune in their lives should be so disproportionately advantaged in their own. And in turn, there is no reason why the imbalances that do exist in who owns capital must also translate into such vast inequalities in how much income people ultimately earn.

Standing between the unavoidable initial imbalances and the ultimate inequality of income is the full spread of institutions that we, as a society, decide to build together: our schools and universities, our taxation and welfare systems, our labor unions and minimum wage laws, to name just a few. These change not only how capital is distributed to begin with, but also what the eventual returns on that capital look like. They determine how this economic prosperity is shared across society.

Inequality, then, is not inevitable. And the same is true for the economic imbalances that technological unemployment would bring about. We have the power to shape and constrain these economic divisions—if we want to.

THE DISTRIBUTION PROBLEM

When Keynes first wrote about technological unemployment, the economic atmosphere was utterly miserable. It was 1930, the Great Depression had begun, and a prolonged period of economic gloom was descending across the entire industrial world. Yet despite the bleakness of the moment, he told his readers not to panic. He asked them to try "disembarrassing" themselves of any "short views" and instead take "wings into the future" with him. In time, he thought, we would solve "the economic problem," the traditional "struggle for subsistence," and we would not need to worry about technological unemployment after all—because by then the economic pie would be large enough for everyone to live on. If technological progress carried on at a steady pace, he calculated, the pie would be the right size within a hundred years—that is, by 2030.[52]

In one sense, Keynes was right in this prediction, and almost a decade before his self-imposed deadline. Today, global GDP per head is almost large enough—as we have seen, it amounts to nearly $11,000 a year—to pull every human being on the planet out of the struggle for subsistence. Anticipating such prosperity, Keynes moved on swiftly in his writing, musing instead on how we might most pleasantly pass our time in this coming "economic bliss."[53] He thought the traditional economic problem of poverty would be replaced by a different problem that was not really about economics at all: how people might put to use all the leisure that technological progress will have won them, how everyone might "live wisely and agreeably and well."[54] For that reason, the prospect of technological unemployment did not worry him at all.

Yet in another sense, Keynes made a serious miscalculation. In his relaxed reflections, he took something significant for granted: that the world's prosperity would automatically be enjoyed by everyone. As we have seen in this chapter, that is far from the truth. As Joseph Stiglitz, a Nobel laureate in economics, put it, "the key issue—to which Keynes repeatedly paid insufficient attention—is that of distribution."[55] Things may look rosy in the global economy as a whole, but "the economic problem" has only been solved for a privileged and lucky few.[56] For most people, their slice of the growing economic pie is still wafer-thin. For many, their share amounts to just some crumbs.[57]

The distribution problem is not new. Inequality has always been with us, and people have always disagreed about how to respond. But the danger is that technological progress will make this distribution problem even more severe, and harder to solve, in the future. Today, many people lack traditional capital, but still earn an income from the work that they do, a return on their human capital. Technological unemployment threatens to dry up this latter stream of income as well, leaving them with nothing at all.

How we should respond is the focus of the rest of the book.

PART III

THE RESPONSE

9

Education and Its Limits

When confronted with the threat of technological unemployment, the most common response from those who think about the future of work—commentators and economists, politicians and policymakers—is that we need more education. From this point of view, the problem facing us is ultimately a skills challenge, and if we give people the right education and training, then this challenge will be resolved. If most people get their income as a return on their human capital, then we have to stop that flow from drying up. Jason Furman, the former chair of President Obama's Council of Economic Advisers, captured this conventional wisdom in a tweet. "Work has a future," he wrote, and "whatever it is, education will help."[1]

For the moment, this is indeed our best response, and the most pressing task that we face is figuring out what "more education" actually means. That is what I try to do in the first part of this chapter. However, as time goes on and machines become ever more capable, education will be of diminishing help. The idea that education can *indefinitely* solve the employment problems created by technological progress is pervasive and largely unchallenged; it is also, as we will see in the second part of this chapter, a big mistake.

THE HUMAN CAPITAL CENTURY

The faith in the power of education to help workers adapt to technological progress comes largely from the past. As we have seen, in the twentieth century technological progress tended to be skill-biased, making the efforts of better-educated workers more valuable relative to those of others. During that time, people who acquired and honed the right skills flourished. And today, an education is still one of the best economic investments a young person could make. If you go to college in the United States, it will cost you about $102,000 on average (in tuition and four years of forgone salary while studying), but as a college graduate you can expect to earn more than $1 million during your lifetime—more than twice the amount you would earn with only a high school diploma.[2] Put another way, a college degree in the United States has an average annual return of more than 15 percent, leaving stocks (about 7 percent) and bonds, gold, and real estate (less than 3 percent) trailing far behind.[3]

Education also does more than just help individuals: it is responsible for thrusting entire economies forward as well. Again, this was particularly true in the twentieth century, so much so that economists called it the "human capital century." During the eighteenth and nineteenth centuries a country's prosperity depended on its willingness to invest in traditional capital, its factories and industrial machines, but in the twentieth century that changed, and prosperity started to depend far more upon an eagerness to invest in human capital, the skills and capabilities of its workers. Why the switch? Because new technologies increasingly required specialized skills, and countries that had better-educated workforces were better equipped to put these technologies to proper use. "Simple literacy and numeracy," write Claudia Goldin and Lawrence Katz, two leading scholars of these changes, were "no longer sufficient" for economic success.[4] More education was needed.

What "more" actually meant, though, changed over the course of the twentieth century. In the beginning, it meant *more people*. The ambition was mass education: that everyone, whatever their background or ability, should have access to proper schooling. This was slow in coming. In the 1930s, Goldin and Katz report, the United States was "virtually alone" in providing free secondary school.[5] But as time passed, other countries caught up and copied that initiative. Today, it is commonplace. And by the end of the twentieth century the meaning of "more" had transformed. It no longer meant simply educating more people, with schooling available to all, but *more advanced* education, with a focus on colleges and universities. You can see this shift in priority in the statements of politicians around the turn of the century. In 1996, US president Bill Clinton introduced sweeping tax changes that he hoped would make "the 13th and 14th years of education as universal to all Americans as the first 12 are today."[6] A few years later, UK prime minister Tony Blair declared that he had "no greater ambition for Britain than to see a steadily rising proportion gain the huge benefits of a university education."[7] And in 2010, President Barack Obama proclaimed that "in the coming decades, a high school diploma is not going to be enough. Folks need a college degree. They need workforce training. They need a higher education."[8]

For now, this is likely to be right, and "more education" remains our best response at the moment to the threat of technological unemployment. But how should "more" be interpreted in the face of increasingly capable machines? The answer involves three changes to our current approach: in what we teach, how we teach it, and when we teach it.

WHAT WE TEACH

In the last few years, an array of policy proposals has come out to respond to the threat of automation. Running through all of them is a single underlying principle: that we must teach people skills that will make them better at whatever machines are bad at, not at what machines are good at. In other words, people should learn to perform tasks in which machines will complement them, rather than substitute for them.

A major implication of this advice is that we must stop teaching people to do "routine" work. As we have seen, "routine" tasks—ones that human beings find it easy to explain how they perform—are where machines already excel, and the substituting force is already displacing human beings. Instead of channeling people into that sort of work, we must prepare them to pursue roles like nursing and caregiving: jobs that involve activities that draw on faculties that, for the moment, remain out of reach of even the most capable machines. Alternatively, we could teach people to build the machines themselves, to design them and set them to appropriate use—another activity that, at the moment, machines cannot really do. For now, focusing on these sorts of activities will give workers the best chance of competing successfully with machines.

Some might bristle at the use of the word *compete*, preferring instead to use one of the many terms that suggest machines help human beings: augment, enhance, empower, cooperate, collaborate. But while words like that may be comforting, they give an inaccurate sense of the changes taking place. Today, new technologies may indeed complement human beings in certain tasks, raising the demand for people to do them; but, as we have seen, that arrangement will only continue as long as people are better placed than machines to do those tasks. Once that changes, though, the helpful complementing force will disappear. The complementing force is only a temporary help: competition, the never-ending struggle to retain the upper hand over machines in any given task, is the permanent phenomenon.

It might be tempting to scoff at the simplicity of the advice here: do not prepare people for tasks that we know machines can do better than human beings. But in fact, that basic principle remains largely ignored in practice. Today we continue to spend a great deal of time teaching people to do exactly the sort of "routine" activities that machines are already superior at, to say nothing about their future capabilities.

Think of the way that we teach and test mathematics, for instance. Many of the problems we set students in high school, if not university, can now be solved by apps like PhotoMath and Socratic: take a photo of the problem, printed or handwritten, with a smartphone, and these apps will scan it, interpret it, and give you an instant answer. It is not a good sign that we still teach and test mathematical material in such a routine way that free off-the-shelf systems like these can handle lots of it with ease. The challenge here is not new: decades ago, basic calculators shifted the emphasis of much mathematical instruction away from brute-force calculation to mathematical reasoning and problem solving. (British students, for instance, take specific exams where calculator use is required and questions are tailored accordingly.) A similar shift is now needed in response to these new technologies as well. And the same principle holds across all subjects: no matter what we teach, we need to explore the material in ways that draw on human faculties that sit out of reach of machines.

On the flip side, right now we also often fail to prepare people for tasks that machines are poorly equipped to do. Take computer science, for example. In the top quarter of US occupations by pay level, almost half of the job postings already require coding skills.[9] In the future, the subject will almost certainly become even more important. And yet in England, for instance, it still remains a dull and uninspiring add-on to the national curriculum, entirely detached from the excitement unfolding at the field's research frontiers. A recent survey found that English computer science teachers often had "no background" in the subject and "do not feel confident" teaching the material.[10] In part, this is because those now tasked with teaching computer science are often the same people who used to teach the old-fashioned (and now discontinued) course in ICT, or Information Communication Technology—where students were taught to use Microsoft Word, Excel, and the like. Policymakers seem to have thought that because both subjects have something to do with computers, there was no urgent need to hire new teachers. Given the quality of the instruction, it is no surprise that only about one in ten students in England take up GCSE computer science at sixteen.[11] More broadly, one in four adults across thirty-three OECD countries has "little or no experience of computers," and most are "at or below the lowest level of proficiency" in using technology to solve problems.[12]

The simple-seeming guidance of preparing people for tasks at which machines will complement them, not substitute for them, is also useful for another reason: it forces us to think more carefully about where exactly in the labor market those complemented tasks are likely to be. Today, it is often assumed that those tasks are found in the most complex and well-paid roles. The aim of much policymaking, therefore, is to encourage people to "up-skill," to pull themselves up through the labor market and try to secure a role at the top. That was the thrust of the comments made by Clinton, Blair, and Obama about college degrees for all. But that twentieth-century strategy is starting to look outdated. As we have seen, the level of education required by a human being to perform a task (in other words, whether it requires a high-skilled worker or not) is a diminishingly useful sign of whether the task can be automated. In fact, many tasks that cannot yet be automated are found not in the best-paid roles, but in jobs like social workers, paramedics, and schoolteachers. Preparing people for such careers will require a very different approach from the traditional one of trying to push an ever-greater number of students through an increasingly advanced formal education.

In the more distant future, however, the simple rule of avoiding "routine" tasks is not going to be enough. We know that machines will not be perpetually confined to "routine" tasks: they are already starting to perform tasks that, in humans, require faculties like creativity, judgment, and empathy. In some ways, machines are also starting to build themselves. (Think of AlphaZero, for instance, the game-playing system that figured out for itself how to become an unbeatable chess player.) This makes it doubtful whether humans eventually will even be able to hold on to work as machine builders.

The trouble with attempting to give any detailed advice for that more distant time, though, is that an impenetrable shroud of uncertainty hangs over exactly *which* tasks will remain out of reach of machines. All we really know with any confidence is that machines will be able to do *more* in the future than they can today. Unfortunately, this is not particularly useful for deciding what people should be learning to do. But that uncertainty is unavoidable. And so we are left with just our simple rule for the moment: do not prepare people for tasks that we know machines can already do better, or activities that we can reasonably predict will be done better by machines very soon.

HOW WE TEACH

Along with changing what material we teach, we also need to change *how* we teach. As many people have noted, if we were able to travel a few centuries back in time and step into a classroom, the setup would look remarkably familiar: a small group of students assembled in a single physical space, addressed by a teacher through a series of live lectures each roughly the same length and pace, following a relatively rigid curriculum.[13] With talented teachers, serious students, and deep pockets to draw from, this traditional approach can work well. But in practice, those resources are often not available, and that traditional approach is creaking.

Today's technology offers alternatives. Take one feature of the traditional approach, the fact that teaching in a classroom is unavoidably "one size fits all." Teachers cannot tailor their material to the specific needs of every student, so in fact the education provided tends to be "one size fits none." This is particularly frustrating because tailored tuition is known to be very effective: an average student who receives one-to-one tuition will tend to outperform 98 percent of ordinary students in a traditional classroom. In education research, this is known as the "two sigma problem"—"two sigma," because that average student is now almost two standard deviations (in mathematical notation, 2σ) ahead of ordinary students in achievement, and a "problem" since an intensive tutoring system like this, although it can achieve impressive outcomes, is prohibitively expensive. "Adaptive" or "personalized" learning systems promise to solve this problem, tailoring what is taught to each student but at a far lower cost than the human alternative.[14]

Or consider another feature of the traditional classroom approach, the fact that there are only a limited number of people who can fit in a traditional classroom or lecture hall before it starts to get too cozy. In contrast, there is no limit to the number of students when teaching is delivered online, no "congestion effects," as economists might say. A computer science course taught by the well-known Stanford computer scientist Sebastian Thrun, for instance, managed to reel in over 314,000 students.[15] There are serious economies of scale with online education, too: the cost of providing a class online is almost the same whether it's seen by a hundred people or a hundred thousand, a pleasing financial situation where the per student cost falls the more students who use the service.[16]

These "massive open online courses," or MOOCs, were greeted with great enthusiasm and fanfare when they first emerged, a decade or so ago. Since then it has become clear that while a vast number of people might sign up for the courses, very few actually finish them; the completion percentage rates are often in the low single figures.[17] We should not be too quick to dismiss this approach, though. Completion rates may be low, but enrollment numbers are very high, and a small proportion of a very large number is often still a large number: for instance, the Georgia Tech online master's in computer science alone boosts the number of Americans with that degree by about 7 percent every year, despite the many dropouts (about twelve hundred Americans enroll annually, and about 60 percent of them finish the program).[18] Also, while students who enroll in MOOCs may not follow through on their initial enthusiasm, its very existence shows that there is a huge demand for education that is currently not being met by our traditional education institutions. This demand can come from remarkably talented places, too. When Sebastian Thrun taught his computer science class to 200 Stanford students, and then to 160,000 non-Stanford students online, the top Stanford student ranked a measly 413th. "My God," cried Thrun on seeing this, "for every great Stanford student, there's 412 amazingly great, even better students in the world."[19]

WHEN WE TEACH

Finally, the third change we need to make in response to increasingly capable machines is how *often* we teach. Today, many people conceive of education as something that you do at the start of life: you put aside time to build up human capital, and then, as you get older, you dip into it and put it to productive use. On this view, education is how you prepare for "real life," what you do to get ready before proper living begins in earnest.

I have been on the receiving end of this way of thinking myself. After working in 10 Downing Street, I returned to academia to study for a postgraduate degree. And when asked at a dinner table what I did for work, I would reply, "I am working on a doctorate in economics." Invariably, my interlocutor would blanch, regretting having steered into a conversational cul-de-sac, and say with a wry smile: "Ah, a *perpetual student*." That response captures an unhelpful conventional wisdom: after a certain age, further education is considered to be a sign not of productivity, but of indolence and flippancy.

In the coming years, this attitude will need to change. People will have to grow comfortable with moving in and out of education, repeatedly, throughout their lives. In part, we will have to constantly reeducate ourselves because technological progress will force us to take on new roles, and we will need to train for them. But we will also need to do it because it is nearly impossible right now to predict exactly what those roles will be. In that sense, embracing lifelong learning is a way of insuring ourselves against the unknowable demands that the working world of the future might make on us.

In some places, these ideas are already ingrained. The Nordic countries, including Denmark, Finland, and Norway, are particularly fond of the idea. And Singapore offers all its citizens over twenty-five a lump-sum credit worth about $370 to spend on retraining, with periodic top-ups to refresh the balance. It is a relatively modest sum, given the scale of the challenge, but distinctly better than nothing at all.[20]

THE BACKLASH AGAINST EDUCATION

If we can adapt what, how, and when we teach, then education is our best current bulwark against technological unemployment. In the last few years, though, there has been a surge of skepticism about the value of education—in particular, about the relevance of the teaching that is currently provided in universities and colleges. Just 16 percent of Americans think a four-year degree prepares students "very well" for a well-paying job.[21] In part, this may have been prompted by the fact that many of today's most successful entrepreneurs dropped out from these sorts of institutions. The list of nongraduates is striking: Sergey Brin and Larry Page left Stanford University; Elon Musk did likewise; Bill Gates and Mark Zuckerberg left Harvard University; Steve Jobs left Reed College; Michael Dell left the University of Texas; Travis Kalanick left the University of California; Evan Williams and Jack Dorsey left the University of Nebraska and New York University, respectively; Larry Ellison left both the University of Illinois and the University of Chicago; Arash Ferdowsi (cofounder of DropBox) left MIT; and Daniel Ek (cofounder of Spotify) left the Royal Institute of Technology.[22]

This list could go on. Though these entrepreneurs stepped away for various reasons, all shared the same trajectory afterward: out of education, and into the stratosphere of the labor market. It is tempting to dismiss them as exceptional cases. It is certainly true that not all dropouts start large, successful technology companies; it is also true that the point of education is not necessarily to raise everyone to start a large technology company. But among those who do, dropouts are not uncommon, and it is a pattern worth reflecting upon for a moment.

Alongside that list's anecdotal power, there are also deeper arguments about why faith in "more education" might be misplaced. The entrepreneur Peter Thiel offers the most provocative version of that case. He claims that higher education is a "bubble," arguing that it is "overpriced" because people do not get "their money's worth" but go to college "simply because that's what everybody's doing." Thiel does not deny that those who are better educated tend to earn more on average, as we saw before. Instead, he is suspicious that we never get to see the counterfactual: how these students would have done without their education. His sense is that many of them would have earned just as much, and that universities are "just good at identifying talented people rather than adding value." Thiel now offers $100,000 grants to young students who choose to "skip or stop out of college" to start companies instead.[23] The Thiel Foundation, which manages the grants, points out that its recipients have started sixty companies worth a combined total of over $1.1 billion. (The foundation omits to mention, though, that we never see their counterfactual, either: what those entrepreneurs would have done without their grants.)

The question of whether universities are "just selecting for talented people who would have done well anyway . . . isn't analyzed very carefully," Thiel complains.[24] In fact, though, many economists have spent large portions of their lives thinking specifically about this issue. The problem is so popular that it has its own name: "ability bias," a particular case of what's known in econometrics as "omitted variable bias." (In this case, the omitted variable is a person's innate ability: if higher-ability people are more likely than others to go to university in the first place, then attributing their greater financial success to their education alone leaves out a significant part of the story.) Economists have developed a tool kit of techniques to address this omission, and their sense—contrary to Thiel's—is that even once ability bias is accounted for, universities still appear to have a positive impact. Talented people might earn more than others in any case, but education helps them earn even more than they would otherwise.

But *how* do colleges and universities help people earn more? There are influential economists—several Nobel laureates among them—who think that it has very little to do with giving students new skills or making them more productive workers. Instead, these economists argue that a large chunk of education is a wasteful phenomenon known as "signaling." In this view, education may well increase people's wages not because it makes them more able, but because it is difficult—so only people who are *already* very able before they start school are able to complete it. So just as a peacock signals his virility to a potential mate by having a particularly fancy set of tail feathers, a student can signal her ability to a potential employer by having a particularly fancy degree. Some suggest that up to 80 percent of the financial reward from education is actually just this ability to stand out from others.[25] On this view, education really has very little to do with giving people new skills at all.

Thiel's general skepticism, then, is important, even if his particular complaint is overstated. Even more important, in general, is a willingness to critique our education system. We tend to treat our schools, universities, colleges, and training centers as if they are sacred: questioning their economic usefulness provokes strong reactions, from discreetly raised eyebrows to more vociferous outrage. Larry Summers, pulling no punches, once described Thiel's $100,000 grants to young students who chose not to go to college as "the single most misdirected bit of philanthropy in the decade."[26] Yet no institution, however venerated and esteemed, should escape critical examination in thinking about the future, our educational institutions included.

THE LIMITS TO EDUCATION

In addition to the current doubts about the value and usefulness of higher education, two other problems are likely to emerge when we look to "more education" as a reprieve from technological unemployment. Of course, education has other purposes beyond simply making sure people are able to find well-paid work, and we will turn to such noneconomic concerns in later chapters. For now, though, I want to focus on education specifically as a response to the economic threat of automation—and on the limitations it has in that regard.

Unattainable Skills

Today, when people propose "more education" as a response to the threat of automation, they do so with a readiness that does not reflect quite how difficult it can be to bring about. New skills are treated as if they were manna from heaven—falling down from the sky in plentiful supply, to be gathered up with little effort by those who need them. But education is not like that. It is hard.[27] It is all very well to say that if workers are displaced by machines, and new work arises that requires different skills, they can swiftly learn them and all will be well. In practice, things do not work like that at all. The difficulty of retraining is part of the reason for frictional technological unemployment: even when there is work out there for human beings to do, those jobs may sit tantalizingly out of reach of people without the skills required to do them. And this is the first limit to education: for many people, certain skills simply may not be attainable.

One reason for this is natural differences. Human beings are born with different bundles of talents and abilities. Some are nimble-footed, others manually dexterous; some have sharp minds, others a finely tuned empathetic touch. These differences mean that some people will inevitably find it easier than others to learn to do new things. And as machines become increasingly capable, narrowing the range of things remaining for people to do, there is no reason to think that everyone will necessarily be able to learn to do whatever is left to be done.

Another reason why skills might be unattainable is that learning to do new things consumes time and effort. We spend large parts of our lives trying to perfect whatever talents and abilities we have, and like the proverbial oil tanker we find it difficult to slow down and change course. Whenever I take a ride in a London taxi, for instance, I am in awe of the drivers: each one has spent years memorizing every street in London, all twenty-five thousand of them, building a legendary body of street smarts known as "the knowledge." Worrying about their future in the age of self-driving cars, I wonder how they might have fared as doctors or lawyers if they had turned their remarkable memories to remembering symptoms and illnesses, or regulations and court cases, rather than destinations and routes. At this point, though, for older drivers a U-turn like this is likely to be a fantasy. What's more, even if it were possible, it might not make financial sense for them. It is one thing to incur the expense of training at the start of your life, with decades of potential earnings ahead to pay it back, but older workers may simply not have enough productive time left in the labor market to recoup it if the burden of repayment falls on them alone.

It would be nice to think that as human beings we are all infinitely malleable, entirely capable of learning whatever it is that is required of us. And you might argue that the difficulty of education is no reason to avoid it. After all, did President Kennedy not say that we do important things "not because they are easy, but because they are hard"?[28] The thrust of Kennedy's comment may be right. But we have to temper our idealism with realism. If "hard" turns out to mean impossible, then inspirational rallying cries to reeducate and retrain are not helpful.

As part of the Program for the International Assessment of Adult Competencies (PIAAC), the OECD recently conducted a survey of literacy, numeracy, and problem-solving skills of adults around the world. The results are striking. "There are no examples of education systems that prepare the vast majority of adults to perform better in the three PIAAC skills areas than the level that computers are close to reproducing," the report states. "Although some education systems do better than others, those differences are not large enough to help most of the population overtake computers with respect to PIAAC skills."[29] In this account, even the *best* existing education systems cannot provide the literacy, numeracy, and problem-solving skills that are required to help the majority of workers compete with today's machines—never mind the capabilities of machines in the future. At present, the survey estimates, only 13 percent of workers use these skills on a daily basis with a proficiency that is clearly higher than that of computers.[30]

Such observations might seem uncompassionate. To highlight differences in ability among human beings feels divisive, and the thought that education might not work for everyone seems pejorative. Moreover, both of these seem to carry an unpleasant undertone that some people are "better" or "worse" than others. In his book *Homo Deus*, the historian Yuval Harari argued that technological trends will lead to the rise of a class of "economically useless people." When he made this point in an interview with Dan Ariely, an influential psychologist, the latter was so irritated and offended that he blurted out: "Don't call them useless!"[31]

Yet Harari's point is not incompatible with Ariely's sympathizing instinct. Harari was arguing, rightly, that some people may cease to be of economic value: unable to put their human capital to productive use, and unable to reeducate themselves to gain other useful skills. He was not claiming that they would end up without any value as human beings. That we so often conflate economic value and human value shows just how important the work that we do (or are seen by others to do) can be. It is a conflation we will return to at the end of this book, when contemplating the search for meaning in a world with less work.

Insufficient Demand

Aside from the difficulty of retraining everyone, the second difficulty with "more education" as an answer to technological unemployment is that it can, at best, only tackle one small part of the problem: the scenario where people lack the skills to do the available work. As we have seen, though, the threat is far more multifarious than that. Frictional technological unemployment is not only caused by workers having the wrong skills: it may also be a product of identity mismatch and place mismatch. (If displaced workers choose not to take up available work because it sits uncomfortably with the type of person they want to be, or if they are unable to move to where new work is being created, then education will not help at all.) But more important, education will also struggle to solve the problem of structural technological unemployment. If there is not enough demand for the work that people are training to do, a world-class education will be of little help.

That is not to say education can be of *no* help in solving the problem of structural technological unemployment. Just as new technologies can increase the demand for the work of human beings by making them more productive at their work, so, too, can education. If doctors or lawyers, for instance, become more productive thanks to better training, they may be able to lower their prices or provide a better-quality service, drawing a bigger clientele. One hope, therefore, is that if structural technological unemployment is caused by a lack of demand for the work of human beings, education could help prop up that demand by making people better at the work that remains for them to do.

As time passes, however, the burden on education to act in this way will grow larger and larger. With technological progress causing the demand for workers to wither away, education will continually have to create more and more demand to make up the shortfall. It is very difficult to see how this could happen indefinitely. As noted before, we are already reaching the point where workers' skill levels are plateauing. There are some limits on how effective education can be in making human beings more productive.

What's more, no comparable limit appears to exist in how productive *machines* could be in the future. As we have seen, when machines operate in different ways from human beings, there is no reason to think that our capabilities must represent the peak of their capabilities. Today, people interested in the future of work spend a great deal of time speculating about the capabilities of machines and where the limits of engineering might be; rarely, though, do we look at ourselves with the same critical eye, to ask about our own boundaries and the limits of education. My sense is that these human limits may be far closer than we think.

THE END OF THE ROAD

When I began to research and write about the future, my preoccupation was with "work." I wanted to know what technological progress would mean for people currently working for a wage: everyone from accountants to bricklayers, teachers to dog walkers, lawyers to gardeners. What would actually happen to them? The reluctant answer I reached is the one set out in the book so far. It is hard to escape the conclusion that we are heading toward a world with less work for people to do. The threat of technological unemployment is real. More troubling still, the traditional response of "more education" is likely to be less and less effective as time rolls on. When I reached this conclusion, my challenge seemed clear: to come up with a different response, one that could be relied upon even in a world with less work.

Yet as I started to imagine what such a response might look like, I came to realize that my focus on the future of *work* alone was far too narrow. Instead, I found myself grappling with the more fundamental question set out in the last chapter: How should we share our society's economic prosperity?

Today, as we have seen, a large part of our answer to that question is "through work." Almost everyone has a bundle of talents and skills, their human capital, and they go out into the world of work looking for a job. In turn, these jobs provide workers with a slice of the economic pie in the form of a wage. This is why we regard work as so vital today, and why the idea of obtaining enough education to keep being employed is so attractive. But it is also why the prospect of a world of less work is so disconcerting: it will put the traditional mechanism for slicing up the economic pie out of use, and make the familiar response of *more education* far less effective than it once was.

Properly responding to technological unemployment, then, means finding new answers to the question of how we share out our prosperity, ones that do not rely on jobs and the labor market at all. To solve the distribution problem in the future, we need a new institution to take the labor market's place. I call it the Big State.

10

The Big State

The great economic dispute of the last century was about how much economic activity should be directed by the state, and how much should be left to the undirected hustle of individuals, free to do their own thing in the market. It was a deep intellectual conflict, a violent clash of ideas regarding the theoretical merits of two very different ways of organizing economic life: central planning on the one hand, and the free market on the other. Friedrich Hayek, perhaps the best-known champion of markets, thought that planning was "the road to serfdom," a path not only to economic catastrophe, but to totalitarianism and political tyranny. Then there were others, like Hayek's student Abba Lerner, who felt quite differently: defecting from his teacher's thinking, Lerner wrote what his biographer described as a "user's manual" for central planners, *The Economics of Control.*[1]

The disagreement divided the world. The United States and its allies thought free markets were the way to go; the Soviet Union and its allies viciously disagreed. At times, central planning appeared to have the upper hand. In 1960, the US government polled ten countries and found that the majority in nine of them thought in a decade's time the Russians would be ahead scientifically and militarily. As the century progressed, statistics trickling out of the Soviet Union painted a picture of astounding economic performance. Then there was the great American humiliation of 1961, when the Soviet cosmonaut Yuri Gagarin became the first person to travel to outer space; hanging in victory above the world, he almost seemed to be mocking the West below. But as the century went on, cracks started to appear—then canyons. We now know the Soviet statistics were not so much massaged as pneumatically drilled into a flattering shape. In the late 1980s, a Russian economist named Grigorii Khanin recalculated the country's growth statistics and published his findings to great outcry in his homeland. While the Soviets had claimed that economic output in 1985 was more than 84 times that of 1928, Khanin found the multiple to be around a measly 7.[2] A few years later, the Soviet Union fell apart.

Given this history, calling for a Big State to solve the distribution problem in the future might sound odd. It appears not just to hark back to this old contest between markets and central planning, but to back the losers of the race—the planners. Didn't the twentieth century emphatically show that they were mistaken? Indeed it did. It provided compelling confirmation that for making the economic pie as big as possible, teams of smart people sitting in government offices and trying to coordinate the economic activity of all citizens according to a master blueprint are no match for the productive chaos of free markets. In calling for a Big State, however, I mean something different: not using the state to make the pie bigger, as the planners tried and failed to do, but rather to make sure that everyone gets a slice. Put another way, the role for the Big State is not in *production* but in *distribution*.

If left to its own devices in a world with insufficient work, the free market—and in particular the labor market—will not be able to continue performing that distribution role.[3] As we have seen, the journey to a world with less work will be characterized by large and growing inequalities. The precedent for dealing with these sorts of major economic imbalances is not encouraging. In the past, such vast inequalities have been reduced on just a few occasions, and only through apocalyptic catastrophes. In Europe, for instance, the last two big falls in inequality were caused by the Black Death plague pandemic in the fourteenth century, and then by the slaughter and destruction of the two world wars in the twentieth. It is hardly a stomach-settling precedent.[4]

This, then, is the reason the Big State has to be big. If we are to find a way to narrow the inequalities by a less cataclysmic route than in the past, it is clear that tinkering and tweaking, as the state has tried before, will not be enough. The only way to deal with the looming disparities is to attack them aggressively and directly.

WHAT OF THE WELFARE STATE?

But do we not already have a Big State—the "welfare state"? It is true that, today, in most of the developed parts of the world, there are many institutions in place alongside the labor market that are designed to support those who find themselves without reliable or sufficient incomes. The particular design, sophistication, and generosity of these mechanisms differs across countries, of course, but they operate in a shared spirit, drawing on a centuries-old argument that says that society has an obligation to help the less fortunate. It is sometimes said that this thinking began with a young Spaniard, Juan Luis Vives, and his 1526 book *On Assistance to the Poor*. At the time, the idea was so controversial that Vives was unwilling to even write the title in a letter to his friend, for fear that "it would fall into the wrong hands."[5] For a long time, the needy relied upon the charity of the prosperous and the free time of volunteers. Gradually, though, local authorities started to respond to beggars and vagabonds by providing them with support or the chance to work.

At the turn of the twentieth century, these welfare institutions started to grow in both generosity and complexity. Countries began to provide unemployment insurance and industrial injury benefits, sickness insurance and old-age pensions, all in an effort to offset the reality that those who lacked a job for any reason would have no income at all.[6] In the UK in particular, serious change started with a 1942 government report called *Social Insurance and Allied Services*, written by economist William Beveridge. Despite its dry title, the Beveridge Report, as it became known, was remarkably influential and well received. Polls showed a majority of all social classes at the time supporting Beveridge's call for more state-provided support. Copies were circulated among troops and dropped behind enemy lines; versions were found, carefully annotated, in Hitler's final bunker.[7]

Since the time of the Beveridge Report, many other proposals have also been made to make sure that everyone in a given society has enough income. Some have remained only theoretical; others became actual policy. For the most part, these plans have tended to piggyback on the labor market, trying to boost people's incomes either by supplementing the wages of low-paying jobs or by attempting to get more people into work in the first place.[8] For example, "working tax credits" or "earned income tax credits" provide tax-offsetting payments to people who earn below a certain amount despite having a job (hence these credits are "earned" through work); most OECD countries have introduced schemes like this in the last few years. Straightforward wage subsidies are another way of addressing insufficient incomes: here, the state, rather than fiddling around with tax credits, instead directly subsidizes low-paid workers to raise their earnings. In various ways, these policies all try to "make work pay"—or, in the case of unemployment benefits, which generally require recipients to be looking for a new job, to make "looking for work" pay instead.

Given that such income-boosting institutions and interventions are already in action around the world, why should we not simply focus on improving and expanding them, perhaps with additional funding and a few tweaks? Why do we need a Big State at all? The answer is that almost all of these schemes were designed for a world where employment is the norm, and unemployment a temporary exception. In a world with less work, neither of these would be true.

Consider the Beveridge Report again, for example. Central to Beveridge's plan for improving British society was the labor market itself. Those who had jobs would make contributions to a collective pot that supported those who could not work (perhaps the ill or the elderly), as well as those who were able to work but found themselves temporarily without a job. The unemployed could draw payments from that pot, but only on the condition that they would be prepared to train for new work while receiving this support. Today, systems like this are often called social safety nets, but they are meant to act more like trampolines, throwing people back up into work after a stumble. Yet if technological unemployment comes about, this approach would fall apart. With fewer jobs, it would be far harder to bounce back after a slip. And the trampoline would start to strain and creak under the weight of all who gathered on it in expectation of support.[9]

The Beveridge Report, with its talk of slaying five "Giant Evils" of society—want, disease, ignorance, squalor, and idleness—did not read like a run-of-the-mill government policy paper. It is furious and polemical, rallying its readers up front with a call to arms (and alms): "a revolutionary moment in the world's history is a time for revolutions, not for patching."[10] Today, we may be approaching a similar moment. Indeed, the challenge we face is probably even larger. The problems of Beveridge's time, though severe, were limited to some segments of society, particularly the poor. But as we have seen, the problem of technological unemployment is unlikely to discriminate in that way. It will reach into many more corners of the labor market. Our instinct should not be to tinker and tweak the institutions we have inherited. Instead, as Beveridge did, we have to free ourselves from old ideas, and be far bolder.

In that spirit, the Big State will have to perform two main roles. It will have to significantly tax those who manage to retain valuable capital and income in the future. And it will have to figure out the best way to share the money that is raised with those who do not.

TAXATION

Today, taxes are not a topic that tends to excite people. People like talking about taxes almost as little as they enjoy paying them. But in a world with less work, taxation will be a critical mechanism in solving the distribution problem. A Big State will have to tax income where it still remains and share it out to the rest of society.

The first question, then, is whom or what to tax. And the simple answer is to follow the income. The previous discussion of the trends in economic inequality provides a strong sense of where money might be found in the future: increasingly, the part of the pie that goes to workers is shrinking relative to the part that goes to those who own traditional capital. What's more, as we have seen, both of those parts are themselves being sliced up ever more unequally, the traditional capital segment particularly so.

As we approach a world with less work, these trends may not hold at the same pace in all places. The Big State will need to be nimble-footed in identifying exactly where income is tending to gather and accumulate. But given current trends, there are three likely places to look.

Taxing Workers

First, the Big State will have to tax workers whose human capital increases in value with technological progress. As we have seen, there will be no "big bang" moment where everyone suddenly finds themselves without work to do. The effects of technological unemployment are likely to be stuttering and uneven. Important, too, there will be some people who escape the harmful effects of task encroachment, who continue to prosper for many years in their work even as others get displaced. New technologies will go on complementing the efforts of these workers, rather than substituting for them. Think of a software developer in the future, for instance, simultaneously more productive because she has increasingly powerful systems at her disposal and more sought-after since the demand for her craft is likely to be voracious. There are others, too, who will likely be able to boost their wages, such as the "supermanager" CEOs we saw before. Both of these types of prospering workers will have to be taxed more than they are now. Economic theory suggests that, even today, the best tax rate to impose on the most prosperous might be as high as 70 percent—quite some distance from where it is at the moment.[11]

Taxing Capital

Second, the Big State will have to tax the owners of traditional capital. This may seem intuitive, given all that has been said about the way that new technologies increase the traditional capital owners' share of the economic pie. But policymakers are still likely to face an uphill struggle, and not just for political reasons. Part of the challenge here will be a theoretical one: according to the most popular models in economics today, the best tax rate to set on this capital is zero. Different models give slightly different reasons for the zero rate. One argument says that capital taxes create distortions that grow explosively over time so they are to be avoided; another essentially proclaims that you can always tax labor efficiently, so why bother taxing capital.[12] And while economists recognize that these models are limited, a feeling has still managed to seep through the economics profession that when talking about taxing traditional capital, the conversation ought to begin close to zero. As Thomas Piketty and Emmanuel Saez delicately put it, "The zero capital tax result remains an important reference point in economics teaching and in policy discussions."[13] This bias will have to be corrected.

A more practical difficulty is that the idea of taxing traditional capital is very ambiguous, far more so than taxing labor. Recently, public discussion has veered toward so-called robot taxes. Bill Gates is partly responsible for this, having caused a stir with his views on the subject. "Right now, the human worker who does, say, $50,000 worth of work in a factory, that income is taxed," he said in a recent interview. "If a robot comes in to do the same thing, you'd think that we'd tax the robot at a similar level."[14]

Well before Gates, others proposed robot taxes as well. Back in the early 1980s, for instance, during a previous bout of automation anxiety, a *Washington Post* reporter found himself in an autoworker union hall "in the sickly heart of car-making country . . . on a slate-gray Sunday afternoon." A union president stood up to declare that "high technology and robots may eliminate each and everyone in this room." As various experts "explained how robots could turn them into blue-collar anachronisms, the auto workers seemed, at first, bewildered. Then they got mad."[15] In response, the Machinists Union drew up a "Workers' Technology Bill of Rights," which, among other things, called for a robot tax. Unemployment, they wrote, would "decrease local, state, and federal revenues" and this "replacement tax" would be needed to fill the gap.[16]

To be sure, there are many problems with the idea of a robot tax. One is that thinking in terms of "robots" is overly simplistic, suggesting that we can perform a simple head count for them as we do for human beings. Even in Gates's simple factory scenario, it is hard to know how to conduct a robot census and what exactly to tax. Another difficulty is that, as we have seen, machines do not simply substitute for workers, but also complement them. Since it is hard to disentangle these effects, how do we know that taxing robots would put a penalty on the harmful ones rather than on the helpful ones? And perhaps most important, we must remember that technological progress (of which robots are a part) drives economic growth—it makes the economic pie bigger in the first place. That is why Larry Summers calls the robot tax "protectionism against progress."[17] A robot tax might mean fewer robots and more workers, but it might also mean a smaller pie as well.

Each of these criticisms has some weight. Together, though, they represent a very narrow interpretation of a "robot tax." If instead we treat the idea in a broader, more charitable way, as just a recognition of the fact that we will need to tax the income that flows to owners of increasingly valuable traditional capital, then it must be right. Quibbles about details and scope do not affect the fundamental point: solving the distribution problem in a world with less work will require the Big State to follow the income, wherever it comes from.[18] In the Age of Labor, most people receive their income as a wage, so human capital has been the most important income source. But in a world with less work, traditional capital will start to matter far more.

A critical first step in taxing traditional capital will be achieving clarity regarding where that capital is located and who actually holds it. At the moment, its location is often unclear. Since the 1970s, the amount of household wealth held offshore, often in tax havens, has shot up; today, it stands at about 10 percent of global GDP, though, as Figure 10.1 shows, there is a lot of variation across countries.[19] Tracing the owners of traditional capital is not an easy exercise, either. Apart from Switzerland, no major financial center publishes detailed statistics on the amount of foreign wealth held by its banks. Many of those who hold this capital do not want others to know about it.[20]

Inheritance tax will also become increasingly important. Of all taxes, this one—often tendentiously nicknamed the "death tax"—ranks as one of the least popular. Parents feel strongly that they should be able to pass on whatever they want to their children; their children feel strongly that they have a right to inherit, unfettered, from their parents.[21] As a result, most places in fact are currently trying to reduce such taxes: in OECD countries, the proportion of government revenue raised by inheritance taxes has fallen by three-fifths since the 1960s, from more than 1 per-

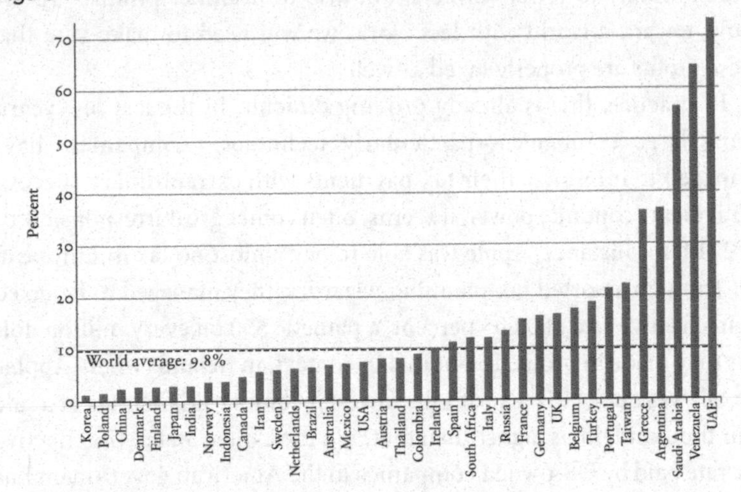

Figure 10.1: Offshore Wealth as a Percent of GDP (2007)[22]

cent to less than 0.5 percent. Some countries have abolished them altogether.[23] This is in spite of the fact that inherited wealth remains a big driver of inequality, and a particularly important explanation for why some people are extraordinarily wealthy. Over the last fifteen years, the number of billionaires in North America who inherited their wealth has increased by 50 percent. In Europe, it has doubled.[24]

In the Age of Labor, we are resigned to the fact that we have to tax people's inherited human capital: when we tax workers' wages and salaries, we indirectly tax the talents they were lucky to be born with, as well as those acquired later on. As we approach a world with less work, we will have to become more comfortable with taxing inherited traditional capital instead.

Taxing Big Business

Third, and related, the Big State will have to tax large companies. In exploring the trends in inequality, we saw that more and more industries are becoming dominated by a shrinking number of corporations. This was important because these superstar firms appear to be partly responsible for the fall in the labor share of income. But this dominance leads not only to fewer workers, but also to healthier profits.[25] As we move toward a world with less work, we will need to make sure that these profits are properly taxed as well.

In practice, this is already proving difficult. In the last few years, many large companies—particularly technology companies—have managed to minimize their tax payments with extraordinary success. With great economic power, it seems, often comes great irresponsibility. In 2014, for instance, Apple was able to pay almost no tax in Europe at all. Through assorted tax-planning wizardry, they managed to be taxed at an effective rate of 0.005 percent, a pathetic $50 on every million dollars they made in profit. To put this in context, in Ireland (where Apple's tax bill was due), citizens with the lowest incomes paid their tax at a rate four thousand times higher than that.[26] In the United States, the effective tax rate paid by US-owned companies to the American government has fallen consistently over the last few decades, even though the nominal tax rate—the one actually set by law—has been steady since the 1990s. Gabriel Zucman, the leading scholar of these trends, estimates that the effective tax rate paid by businesses to the US government fell by 10 percent between 1998 and 2013, with about two-thirds of that reduction due to increased tax avoidance. Again, tax havens play an important role in enabling this: since 1984, the share of US corporate profits reported in places like the Netherlands, Luxembourg, Ireland, Bermuda, Singapore, and Switzerland has increased more than eightfold.[27]

Tax avoidance by large corporations usually does not break the letter of the law governing corporate taxation, though it frequently offends the spirit of the law. (Sometimes it does both: the European Commission determined that Apple's Irish tax breaks violate international regulations, and served Apple a bill of €13.1 billion.[28]) In other words, corporate tax avoidance stirs public outrage not because it is illegal but because it is immoral.[29] When highly profitable businesses rely on legal loopholes and technicalities to avoid paying a reasonable level of tax, it is seen as a betrayal of the trust that people placed in these companies.

This does not mean that people should be thrilled to pay their taxes. Not everyone has to share the enthusiasm of the great American jurist Oliver Wendell Holmes, who once said, "I like to pay taxes. With them I buy civilization."[30] What it does mean, though, is that there is a need

Figure 10.2: Nominal and Effective Tax Rates on US Corporate Profits[31]

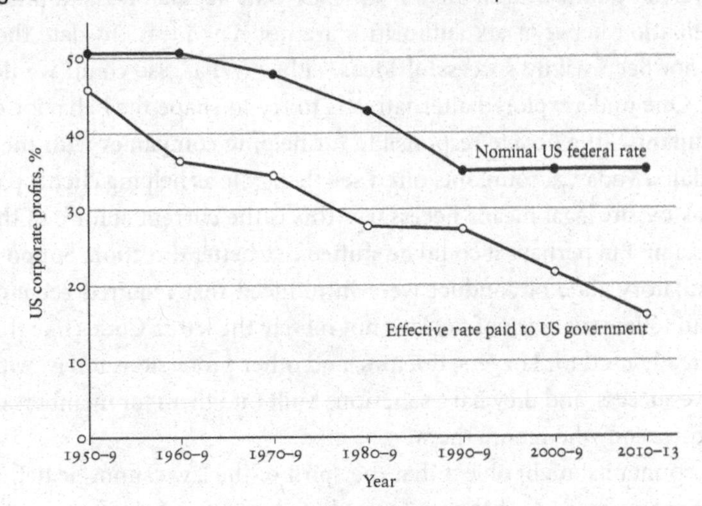

to bring the letter of corporate tax law in line with the spirit. In other words, we need tighter legislation so that big companies are forced to pay their fair share.

The next challenge will then be to enforce any such new legislation. In part, this requires greater political will and grit than what we see today; for instance, the number of potential tax cheats that the US Internal Revenue Service refers for criminal prosecution, relative to the size of the population, has collapsed by 75 percent over the last twenty-five years.[32] In part, it also requires better officials and regulators—or, at least, officials and regulators who are as capable as the companies they are regulating. Most major companies are likely to instruct their tax experts to devise clever new moves and uncover new loopholes to avoid whatever rules are imposed on them. There is also the problem that even if regulators do catch up to the tax advisers, and one country is able to enforce a higher effective tax, a company can simply slip away and relocate somewhere else where taxes are lower. Many businesses, rather than bother to evade stricter tax laws, will move instead. (This is true not only for large companies, of course, but all owners of valuable capital.) For that reason, better international coordination to prevent such tax-evading relocations will be needed, too.

Greater political will, better officials and regulators, and more coordination between tax authorities are not new ideas. To date they have not been wildly successful ideas, either. What else could we do, then? One underexplored alternative is to try to shape the behavior of accountants—the people responsible for helping companies with their tax affairs. Today, accountants often see their role as helping clients pay less tax by any legal means necessary. This is the current culture of the profession. But perhaps it could be shifted in a better direction. Suppose a mandatory code of conduct were introduced that required accountants to follow the *spirit* of tax law, not merely the letter. Codes like this are already used for lawyers, doctors, and other professions today, with relative success, and they have sanctions built into them for members of the profession who ignore them.

Accountants might object that the spirit of the law cannot be definitively ascertained. And that is true—but the letter of the law can be mind-bogglingly ambiguous, too. That is how accountants make their money, after all: by helping people navigate and exploit those uncertainties in the tax system. Under this new code of conduct, their work would shift to helping navigate ambiguities in the spirit of the law instead, a step toward ending the tax avoidance industry that flourishes today.

THE INCOME-SHARING STATE

Once the Big State has raised the necessary revenue, the next question is how to share it so that everyone has enough income. In the twentieth century, as we have seen, answers to this tended to rely on the labor market. Revenue was spent on raising the wages of the lowest-paid workers and on supporting those who found themselves unemployed while encouraging them back into the job market. In a world with less work, however, these approaches will be markedly less effective than they were in the past.

This is why, among those who worry about the future of work, there is a lot of excitement about the idea of a universal basic income, or UBI. This scheme sidesteps the labor market altogether: it is a regular payment that the government provides to everyone, whether or not they are employed. Support for the UBI can also be found well beyond just those who are anxious about automation: it is one of those rare policy proposals that makes the political spectrum bend back on itself, with people on opposite ends meeting in violent agreement. Conservatives like the UBI because it is simple, promising to do away with the inefficient complexity of existing welfare systems, while liberals like it because it is generous, promising to get rid of poverty altogether. For our purposes, though, it is its implications for the world of work that are particularly interesting.

This enthusiasm for the UBI may be relatively new, but the idea itself is not. It was first proposed by Thomas Paine, one of America's founding fathers, who published a pamphlet about it in 1796. In the opening pages, Paine describes his irritation on hearing a bishop preach that "God made *rich* and *poor*." This, he thought, was completely wrong. God did not create inequality, Paine argued, but gave everyone "the earth for their inheritance" to share. And yet, looking around him, Paine saw that in practice only a small class of landowners came to enjoy that inheritance. To solve this, he proposed that everyone should be given an annual lump of cash to compensate them for their loss—a sort of UBI.[33] Since then, the idea has continued to appear in various guises: as a "territorial dividend" and "universal benefit," "citizen's income" and "citizen's wage," "state bonus" and "demogrant." (Today's preferred label, "basic income," appeared in the twentieth century.) Along the way, the concept has gathered illustrious supporters, from Bertrand Russell to Martin Luther King Jr.

The wide range of support for the UBI disguises the fact that key details of it are subject to uncertainty and disagreement. For instance, how are payments made? UBI supporters often argue that payment in cash is a "fundamental" part of their proposal, but in practice there are other reasonable ways to make people more prosperous.[34] One approach, for instance, is to make important things available in society at no cost: rather than just give people cash, the state in effect makes certain purchases on their behalf. Already in the United States, about forty million people use the Supplemental Nutrition Assistance Program, or "food stamps," to receive basic sustenance for free, worth about $1,500 a year.[35] In England, health care and primary and secondary education are free for everyone who wants them, each worth thousands of pounds per year.[36] Add up such initiatives, and you end up with a sort of UBI—though one that the state has already spent for you.

And if the income payments do get made in cash, how generous should they be? The UBI says "basic." But what does that mean? Some economists think it implies a minimal payment, not very much at all. John Kenneth Galbraith, for instance, said that introducing "a minimum income essential for decency and comfort" is the right thing to do.[37] Friedrich Hayek similarly spoke of "a certain minimum income for everyone."[38] Today's prominent UBI advocates often agree. Annie Lowrey, author of *Give People Money*, makes the case for "just enough to live on and not more"; Chris Hughes, author of *Fair Shot*, argues for $500 a month.[39] But there are others who feel differently. Philippe Van Parijs, today's leading UBI scholar, wants to use UBI to build a "truly free" society, where people are not tied down by what they earn. That is a far loftier goal than what is envisaged by Galbraith and Hayek—and a far more expensive one, too. Or consider Thomas Paine, the man who invented the idea in the first place. For him, it was not about alleviating poverty, like Galbraith, nor providing security, like Hayek, nor achieving freedom, like Van Parijs; it was about compensation for lost farmland. Paine wanted the UBI to be large enough for everyone to "buy a cow, and implements to cultivate a few acres of land"—worth, it is said, about half the annual earnings of a farm laborer at the time.[40] Again, that adds up to a rather sizable sum.

In large part, then, how basic "basic" actually is will depend on what the payment is intended to do. For Galbraith and Hayek, the emphasis was on a "minimum" because their ambitions for a UBI were relatively modest. Galbraith envisioned his UBI only as a floor in the standard of living beneath which nobody should be allowed to fall. Hayek proposed it to make sure that people had a baseline level of economic security: "food, shelter, and clothing" so everyone was fit and healthy and able to work, and little else. But if we are thinking about a basic income in the context of a world with insufficient work, the aim is likely to be much closer to the ambitious goals of Van Parijs and Paine. In that world, for many people the payments would provide not just a baseline income that they could top up through their work, like Hayek or Galbraith imagined, but their entire income full stop.

Finally, there is one more question to ask about the idea of a basic income. What are the conditions attached to the payment? Most UBI proponents would answer that, by definition, there are none. But in a world with less work, I believe, it is crucial to depart from this assumption. To deal with technological unemployment, we will need what I call a *conditional* basic income—a CBI, for short.

A CONDITIONAL BASIC INCOME

People who say that a UBI should be "universal" tend to have two things in mind: the payment is available to everyone who wants it, and it is made available without imposing any requirements on the recipient. The CBI I propose is different in both of those respects. It is only available to some people, and it explicitly comes with strings attached.

The Admissions Policy

When UBI advocates say that the basic income payment should be made available to everyone, most of them don't really mean *everyone*. A literal interpretation of universality would imply that any visitor could drop by a country with a UBI, pick up their payment, and head right back home with a fatter wallet. To avoid this scenario, most advocates imagine that a UBI would only be available to the citizens of the country paying it out. (That is why it is sometimes called a "citizen's income.") This adjustment is often treated as if it is the end of the matter. In fact, though, it is just the beginning. A fundamental question remains unanswered: who gets to call themselves a citizen? Who is in, and who is out of the community? In a UBI, the *admissions policy* is missing.[41]

Over the last few decades, Native American tribes have shown just how contentious drawing up an admissions policy for a community can be. Within the borders of Native American reservations, the tribes have a degree of "tribal sovereignty," meaning they are allowed to run some of their own affairs.[42] Economic life on the reservations has always been very difficult: Native Americans have the highest poverty rate of any race group (26.2 percent compared to the US national average of 14 percent), and a suicide rate among young people that is 1.5 times the national average.[43] In response, some reservations have used their sovereignty to start up gambling operations, building casinos to lure outsiders onto the reservations and boost the local economy. Today almost half of the tribes run casinos, some very small, but others large enough to rival the grand spectacles of Las Vegas. It is big business, too—over $30 billion in annual revenue.[44]

Some of the successful tribes, flush with income, have drawn up "revenue allocation plans" to share out this money among their members. And these plans look a lot like a UBI: all members of the tribe, often through no real productive effort of their own, get a slice of the income. The sums can be vast, as much as several hundreds of thousands of dollars per person every year. But here is the problem: payments like this create a huge economic incentive for recipients to kick others out of the group in order to secure a bigger slice of the income for themselves. And this is exactly what is happening with Native Americans, as long-standing community members find themselves getting expelled from their tribes by corrupt tribal leaders.

As we approach a world with less work, this sort of struggle over who counts as a member of the community will intensify. The Native American experience shows that dealing with questions of citizenship is likely to be fractious. The instinct in some tribes was to pull up the drawbridge—a reaction we can see in other settings, too. Consider the financial crisis in 2007 and its aftermath. As economic life got harder, the rhetoric toward immigrants in many countries hardened as well: they were said to be "taking *our* jobs," "running down *our* public services." There was a collective impulse to narrow the boundaries of the community, to restrict membership, to tighten the meaning of *ours*. In much the same way, support for so-called welfare chauvinism—a more generous welfare state, made available to fewer people—is on the rise. In Europe, for example, a survey found "both rising support for redistribution for 'natives' and sharp opposition to migration and automatic access to benefits for new arrivals."[45]

In the Age of Labor, there has been a persuasive economic response to this instinct to exclude others: through their work, immigrants make the country's economic pie bigger. As a result, letting in more people does not necessarily leave existing citizens with a smaller slice; on the contrary, there is often more income per capita to share out. But in a world with less work, that response will be far less compelling. There will be fewer opportunities for newcomers to contribute through their jobs, and a greater chance that they will depend on the efforts of others for their income. In that world, it is more likely that adding new members to a community will in fact lead to existing members having smaller slices of the pie. At that point, it will be far harder to respond to the hostility to outsiders with the economic reasoning of the past.

In short, the world of less work will not let us avoid the question of who is in the community and who is out. A CBI will force us to address that issue directly, rather than trying to dodge it with a UBI.

The Membership Requirements

The second piece of what UBI advocates mean when they say that payment is "universal" is that there are no demands placed on those who receive it. Whether a person is in work and what they earn, for instance, does not matter. There are no "means tests" or "work tests," no strings attached. Put another way, once people meet the admissions policy, there are no *membership requirements* for maintaining eligibility.

This is sometimes a source of puzzlement. It seems to imply that a UBI would go not only to those with a very small income, who might really need it, but also to those with vast incomes, who do not. That appears to be a very poorly targeted waste of money. UBI advocates tend to reply that, on the contrary, it is actually very important that everyone receives the payments. First, they say, it is not wasteful: if payments are funded through taxes, then the rich may receive a payment, but they will also pay far higher taxes to support other people's payments, more than making up for the income they get. Second, the approach makes practical sense: universal payments are easier to administer and less confusing for the recipients, removing any uncertainty about eligibility. And third, and most important, UBI advocates argue that universal payments remove any stigma associated with claiming support. If everyone receives the payments, nobody can be labeled by society as a "scrounger" and no individual will feel ashamed to have to claim theirs. As Van Parijs puts it, "There is nothing humiliating about benefits given to all as a matter of citizenship."[46]

The idea of payments made without any conditions also runs completely contrary to how things tend to be done today. Most supportive payments from the state do tend to come with strict requirements, often demanding that recipients are in work (albeit work that might not pay very much) or vigorously seeking employment. This is partly because economists worry that, without such strict requirements, state payments will create a strong disincentive to work—encouraging those in work to work less, and those without work to stay put. Imagine somebody on the fence about the labor market, not sure whether to get a job or not. A guaranteed income paid to them regardless of what they choose to do may well tip them off the fence and out of the market. As it happens, the evidence on whether a lack of requirements actually creates these disincentives is not very decisive.[47] Nonetheless, some economists are suspicious that a no-strings-attached UBI will harm its recipients' willingness to work.

In today's Age of Labor, these disincentive effects may be a sensible reason to think that a basic income must come with some conditions: you want to make sure that those who receive it still want to work. However, as we move toward a world with less work, this argument becomes far less compelling. Encouraging people to work only makes sense if there is work for everyone to do, which would no longer be the case.

There is a different reason, however, for why a basic income in a world with less work should have conditions attached—why it should be a CBI rather than a UBI. The point will not be to support the *labor market*, but to support the *community* instead.

A world with less work will be a deeply divided one. Many members of the community will not be able to make much of an economic contribution of their own, and instead will have to rely on the productive effort of others for their income. Keeping such a split society together will be a serious challenge. How do you make sure that those who receive payments, but do not work for them, are thought to deserve them? How do you avoid feelings of shame and resentment on either side? After all, those are not unprecedented reactions. They already exist in today's welfare state—and, somewhat ominously, they crop up in response to payments that are far more modest than those that will be required in the future.

The UBI fails to take account of these responses. It solves the *distribution* problem, providing a way to share out material prosperity more evenly; but it ignores this *contribution* problem, the need to make sure that everyone feels their fellow citizens are in some way giving back to society. As the political theorist Jon Elster put it, the UBI "goes against a widely accepted notion of justice: it is unfair for able-bodied people to live off the labor of others. Most workers would, correctly in my opinion, see the proposal as a recipe for exploitation of the industrious by the lazy."[48]

In contrast to the UBI, today's labor market tackles both of these problems at the same time. It solves the distribution problem, at least to some extent, by paying people a wage in return for their work. And it addresses the contribution problem by allowing people to contribute to the collective pot through the work that they do and the taxes they pay. Social solidarity at the moment comes, in part, from the fact that everyone who can is trying to pull their economic weight. Unpleasant rhetoric gets directed at those who do not—think of the tendency to label those who rely on others as "spongers" and "free riders."

In a world with less work, it will no longer be possible to rely on the labor market to solve the distribution problem, as we have seen—or this contribution problem, either. So how can we re-create that sense of communal solidarity? A big part of the answer must involve membership requirements attached to the basic income. If some people are not able to contribute through the work that they do, then they will be required to do something else for the community instead; if they cannot make an economic contribution, they will be asked to make a noneconomic one in its place. We can speculate about what these tasks might turn out to be; perhaps certain types of intellectual and cultural toil, caring for and supporting fellow human beings, teaching children how to flourish in the world. It will fall to individual societies to settle on what these contributions should look like, a theme we will come back to in the final chapter.

The Question of Diversity

Another part of the answer to the question of communal solidarity might involve making the CBI's admission policy more exclusive. There is a large, albeit contested, body of research that suggests a trade-off between the diversity of a community and the generosity of state provisions to it. Economists have found that US cities that are more ethnically fragmented, for instance, tend to have lower spending on public goods, like education and roads, sewers and garbage collections.[49]

There is an argument that race might explain why America lacks the sort of generous welfare state you find in many European countries. Racial minorities are highly overrepresented among impoverished Americans, and due to poor race relations, other Americans might be unwilling to support a generous welfare state that would disproportionately help those minorities.[50] The political scientist Robert Putnam caused controversy with a study showing that inhabitants of diverse communities are less likely to trust anybody. "In the presence of diversity, we hunker down," he said. "It's not just that we don't trust people who are not like us. In diverse communities, we don't trust people who do look like us."[51]

Clearly, such findings should not be celebrated. If they are right, the way to improve welfare provision in the United States is to enhance race relations, not to push for a more homogenous population. Putnam, for his part, was furious when his research was "twisted" by other academics to make the case for less diversity.[52] His overall message, he said, was one of inclusivity, not exclusivity: that we ought to build a bigger sense of "we" to fight against discord and distrust.

Nevertheless, these results should prompt broader questioning. Diversity is not only about race. Most of us do feel stronger obligations to our families than to a stranger on the other side of the globe; and somewhere between those two extremes sit our communities of shared places and mutual interests, similar work and common country. What is the moral significance of these communities? Is valuing or protecting them necessarily parochial and xenophobic? As the political philosopher Michael Sandel asks, are there no "legitimate grievances" buried in the frustration of those today who chant, "American Jobs for American

Workers"?[53] What about the equivalent slogan in a world with a basic income—"American Incomes for American Citizens"? And even if you think that such communities have no *moral* significance, what about their *practical* significance? What if tightening membership criteria to strengthen social solidarity is the only way to stop a community with cavernous economic divides from falling apart?

In the century to come, questions about distributive justice, how we share out resources in society, will become more urgent. But these questions about contributive justice, how we make sure that everyone feels their fellow citizens are giving back to society, will become more pressing as well. A UBI engages with the first set of questions, but not the second. The CBI, by explicitly confronting who is eligible for the payment and on what terms, addresses both.[54]

THE CAPITAL-SHARING STATE

The primary role for the Big State, then, will be to tax and share out income, perhaps with a new set of noneconomic conditions attached to build social solidarity. But there are also other things that it could do to help address a world with less work. One of these is to share out the valuable capital itself, the source of that income in the first place. Whereas a UBI or CBI provides a basic income, this would be a basic *endowment*—giving people not a regular flow of cash, but their own stock of traditional capital to hold on to.[55]

There are two reasons why sharing out capital might be attractive. The first is that it would reduce the need for the Big State to act as an income-sharing state. If more people owned valuable capital, income would flow more evenly across society of its own accord. The second reason is that such sharing would also help to narrow economic divisions in society. If the underlying distribution of capital stays the same, and the state only shares out income, then profound economic imbalances will remain. If left unresolved, such divisions could turn into noneconomic strife: ruptures of class and power, differences in status and respect.[56] By sharing out valuable capital, and directly attacking the economic imbalances, the state could try to stop this from happening.

In a sense, governments already do this: since the start of the twentieth century, the state has tried to share *human* capital as widely as possible. This is the point of mass education. Making good schools and universities open to all is an attempt to make sure valuable skills are not left only in the hands of a privileged, well-educated few. Now, as we move out of the Age of Labor, the Big State must try to share traditional capital as well.

The sharing of traditional capital could happen without the state having to step in—but it is very unlikely. For one cautionary tale, consider a company called Juno. Like Uber, Juno is a ride-hail company, but with an important difference: while Uber was owned by its founders, Juno was initially owned by some of its drivers as well. When drivers joined Juno, they were given the chance to receive some Juno stock, which could translate into a stream of income if the company came to enjoy financial success. That was the promise, but it was never kept. A year after it was founded, Juno was bought by Gett, another taxi company, whose owners promptly voided the drivers' stock plan. The new owners could not resist taking control of the valuable capital and keeping the income for themselves.[57] The fact that Juno's initial plans were so widely celebrated, and that other examples of such arrangements are so rare, suggests that the free market alone is unlikely to share traditional capital out by itself.

In theory, buying shares in the stock market gives people the same opportunity to acquire an ownership stake in many companies. The problem, of course, is that the stock market, as has been said of the legal system, "is open to all—like the Ritz Hotel."[58] In practice, most people do not have either the financial wherewithal or the know-how to invest profitably in this way. In the United States, for instance, reflecting the inequalities we saw earlier in the book, almost everyone in the top 10 percent of earners own stocks, but only about a third of the people in the bottom 50 percent of earners do.[59] One possibility, then, is that a capital-sharing state might acquire a stake on behalf of those without one, pooling their investments into a fund on behalf of its citizens—a Citizens' Wealth Fund.

There is a precedent for this. Today's sovereign wealth funds, large pools of state-owned wealth that are put into a spread of assorted investments, perform a similar role. The world's largest such fund, worth over $1 trillion, is owned by Norway. After Norway started developing its oil reserves, rather than spend all the profits right away, the government set up a fund "on behalf of the Norwegian people."[60] Since the country has a population of about 5.2 million, each citizen effectively has a stake worth about $190,000. Every year, some of the fund is siphoned off and spent on the Norwegian economy.

Then there is the Alaska Permanent Fund, worth a more modest $60 billion. Since 1976, about a quarter of the annual royalties from the production of oil and gas in the state of Alaska have been saved in a fund. And each year, a percentage of the fund is also siphoned off and spent on all Alaskans, in this case via direct payments to every resident of the state—about $1,400 a year for every adult and child.[61]

At the moment, though, funds like these are exceptions. In many countries, as shown in Figure 10.3, the amount of publicly held traditional capital relative to the size of the national economy is falling, whereas the amount of privately held traditional capital relative to the size of the national economy is rising.

James Meade, the Nobel Prize–winning economist, anticipated the capital-sharing role for the Big State back in the 1960s. Worrying about automation in the future, he suggested a possible approach where the

Figure 10.3: Private Capital and Public Capital[62]

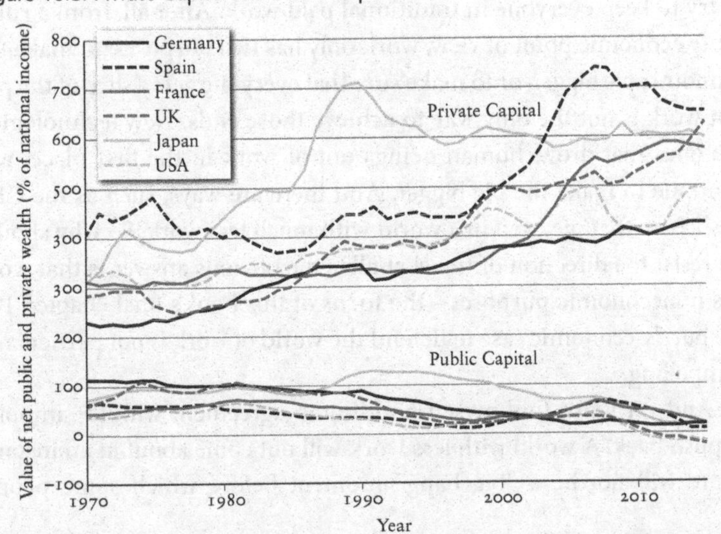

state would own capital on everyone's behalf. He called this the "socialist state," but that label is not quite right: it ignores the difference between the state having some *ownership* in these companies, which is what I have in mind, and the state having complete *control* over how these companies operate, which is what traditional socialists tend to be preoccupied with. As a name, the "capital-sharing state" is a better fit.

THE LABOR-SUPPORTING STATE

Until now, I have taken the direction of economic travel for granted—toward a world with less work. With that in mind, the role I have set out for the Big State is to go with that flow, to step in and redistribute the economic prosperity created by these new technologies if the labor market, our traditional mechanism for sharing out prosperity, can no longer be relied upon to do it. There is an alternative, though: to resist this direction of travel. A Big State would still be needed, but to work in the opposite direction—not to be passively carried along on the current of technological progress, but to actively defend the world of work from the changes that are unfolding.

My response to this notion is to shake my head and nod simultaneously. The economist in me sees little reason to defend the labor market, to try to keep everyone in traditional paid work. After all, from a ruthlessly economic point of view, work only has two purposes: to make the economic pie bigger, or to make sure that everyone gets a slice of the pie. But work is not the only way to achieve those ends. New technologies, the ones that drove human beings out of work in the first place, will continue to make the pie bigger. And there are ways, such as the CBI, to slice up that pie even in a world with much less work. So why should we resist the direction of travel at all? One obvious answer is that work has noneconomic purposes—the focus of this book's final chapter. But the purely economic case to defend the world of work is not particularly compelling.

And yet I also find myself nodding in agreement with the impulse to push back. A world with less work will not come about in an instant. There will not be a "big bang" moment before which some people have work and after which they do not. Instead, there will be a stuttering decline in the demand for the work of human beings, beginning in small corners of the labor market and spreading as time goes on. While this is happening, the world of work will see changes not only in the number of jobs, but in the pay and quality of those jobs as well. In turn, as the demand for human beings gradually falls away, workers will wield a weaker economic punch, carry less economic clout, and possess a diminished collective ability to challenge profit-seeking employers, whose incentive will tend to be to pay them as little as they can. Workers do not start from a strong position today, either: in the developed world, organized labor has slumped over the past decades, and trade union membership has nose-dived.[63]

This means that the Big State has a middle ground between trying to change our direction of travel and passively going along with it. It can be the labor-supporting state: stepping in to support workers during this transition, to make sure that whatever jobs remain are well-paid and high-quality. The purpose is not to change our final destination, but to make sure the journey is as smooth for workers as it can be. So long as there is work to be done, there will be a role for the state in making sure it is "good" work, especially since workers acting alone will have an ever weaker ability to do this for themselves. John Kenneth Galbraith coined the term "countervailing power" to describe the different forces that might hold concentrations of economic power in check.[64] In the twenty-first century, as the countervailing power wielded by workers falls away, the state should step in to wield it on their behalf.

We have to be realistic about the way in which a labor-supporting state can work. There is a popular line of thinking that says we ought to ask businesses to develop new technologies that complement rather than substitute for human beings, that help rather than harm workers. Microsoft's CEO, Satya Nadella, has called this "the grand challenge."[65] But simply asking companies to do this, if it is not in their financial interest, is akin to asking them for charity—an idealistic, unrealistic basis for large-scale institutional reform. At the 2019 World Economic Forum in Davos, the *New York Times* reports, business leaders talked a good game in public about how to contain "the negative consequences that artificial intelligence and automation could have for workers," but in private these executives had "a different story: they are racing to automate their own work forces to stay ahead of the competition."[66]

In trying to shape how institutions act, we must do so on the basis of how people actually behave. We have to take them as they are in economic life, selfish and partial, not how we would like them to be, benevolent and impartial. For that reason, the state's labor-supporting efforts should be focused primarily on changing the actual incentives that employers face, forcing closer alignment between their interests and those of the society of which they are a part.

One way to align employers and society is through the tax system. In the United States, for instance, the tax system right now unwittingly encourages automation, granting those who replace human beings with machines "several major tax advantages"—such as not having to pay payroll taxes on employee wages.[67] The problem is that this system was created in the Age of Labor, and so was designed to raise revenue in large part by taxing employers and employees. It was not designed to operate in a world with less work. Removing such advantages will get rid of this incentive to automate.

Another approach is to use the law. For instance, there is an ongoing controversy over the legal status of Uber drivers. Are they self-employed, as Uber argues, left to fend for themselves as they run their own driving businesses? Or are they Uber employees, entitled to holiday pay, a pension, the minimum wage, and all the other rights that come with that status? There is a role here for the labor-supporting state to help these workers, by updating the law to bring them under protections akin to those offered to people elsewhere in the labor market. Other legislative interventions, for instance, could set new floors beneath which the level of pay is not allowed to fall, building on established minimum wage regulations.

There is scope here to be innovative. Traditionally, policymakers set minimum wages with living standards in mind, trying to make sure that the lowest-paid workers still have enough to live on. But there are other criteria that could be used in addition to that. For instance, a distinctive feature of many hard-to-automate roles, like caregiving and teaching, is the vast gap between their economic value and their social value: this work tends to be low-paid, and yet is widely recognized as being extremely important. In the UK, a poll found that 68 percent of people think nurses are underpaid; in the United States, 66 percent think that public school teacher salaries are too low.[68] When the labor-supporting state intervenes to influence wages, it could take the opportunity to narrow this gap as well.

In a similar way, policymakers traditionally think about working-time regulations in terms of hours. In Europe, by law, your employer cannot force you to work more than forty-eight hours a week. Some countries aspire to forty hours. Germany's largest union even secured a twenty-eight-hour working week for its members in 2018 (and a 4.3 percent pay rise along with it).[69] In time, though, it may make sense to establish limits not just on the number of *hours* but also *days* per week. In 2018, for instance, the Trades Union Congress (TUC), the collective voice for forty-eight unions and their 5.5 million members in the UK, called for a four-day workweek as a possible response to automation.[70] This is the sort of proposal that should be taken increasingly seriously.

A final role for the labor-supporting state is blunter: to encourage the emergence of new forms of organized labor. In the twenty-first century, trade unions must not only help workers respond to technological change, but must also use that very same technology to transform the way that they work. At the moment, the ways in which unions recruit members, raise finance, express grievances, and exercise their power often look remarkably similar to the antiquated methods that have been used for centuries. Few trade unions, if any, provide their members with access to customized e-mediation platforms or dispute resolution systems, despite the success of such systems elsewhere. Social networks and digital tools remain peripheral to old-fashioned ways of working; the rise of "connective action," where people use technologies to coordinate and cooperate, is largely happening outside traditional unions.[71] In part, this explains why membership has fallen so precipitously among younger people: they simply do not see today's unions as a relevant response to contemporary problems. In the UK, less than 8 percent of workers aged sixteen to twenty-four are members. (Among those who are members, 40 percent are aged fifty or over.)[72] Frances O'Grady, the leader of the TUC, recognizes the challenge; "Unions have to change too—change or die," she admits.[73]

Just as technological unemployment will not happen overnight, there is no need for the Big State to establish itself in the coming weeks. But as time goes on, the need for one will only grow. Some combination of the three roles (the income-sharing state, the capital-sharing state, the labor-supporting state) will eventually be required to keep our increasingly divided societies from falling apart. This chapter is not meant to be overly prescriptive about what these roles involve. There is no definitive list of interventions that all countries must adopt. There is a variety of different ways in which the Big State could deal with technological unemployment—and it must fall to citizens in each country, to their unique moral tastes and political preferences, to determine how exactly to strike the best balance between them.

11

Big Tech

As we approach a world with less work, our economic lives will become increasingly dominated by large technology companies. And with that growing economic power the companies will acquire great political power, too. They will not only shape how we interact in the marketplace, what we buy and sell, but how we live together in broader society, our existence as political animals as well. Understanding the rise of Big Tech and the nature of their growing political power is as important as making sense of the decline of work—for in a world with less work, constraining these companies will demand more and more of our attention. The challenge is that, at present, we are almost entirely unprepared to respond effectively.

WHY TECH?

Today, when we think about technology companies, the "Big Five" come to mind: Amazon, Apple, Google, Facebook, and Microsoft. And their figures are startling. In the United States, Google gets 62.6 percent of search engine traffic and has 88 percent of the market in search advertising.[1] Facebook is used by almost a third of the world's human beings, and across its various platforms (like Instagram and WhatsApp) it is the gatekeeper for 77 percent of mobile social traffic. Amazon is the shopkeeper of choice for 43 percent of all online retail and 74 percent of the e-book market.[2] Apple and Google combine to control 99 percent of mobile phone operating systems. Apple and Microsoft account for 95 percent of desktop operating systems.[3] In 2018, these five companies were among the top-ten most valuable in the world.[4]

Notwithstanding these remarkable numbers, though, we should not be too preoccupied specifically with this small collection of names. Yes, those companies are likely to remain prominent for some time. But new technologies that reshape our lives will also come from people and institutions well beyond the Big Five. Indeed, point to any part of modern life, and you can be fairly certain that someone, somewhere, is working away in a metaphorical garage, trying to develop a new system or machine to change it. In 2011, the venture capitalist Marc Andreessen wrote that "software is eating the world."[5] In the years since, its appetite has indeed proven voracious. There are very few industries, if any, that new technologies do not find at least partially digestible. All corners of our lives are becoming increasingly digitized; and on top of our world of physical things we are building a parallel world of ones and zeros. In the future, it is hard to see how our economy will escape being run almost entirely by technology companies of various sorts.

Of course, some of these dominant powers might be the same technology companies we already know. When IBM developed Deep Blue, and Google acquired DeepMind, they did not do so because they wanted to win at board games. They spent their respective fortunes in the pursuit of far bigger ambitions—at times, astonishingly grand ones. The story of WeChat is an inspiration to them. What began as a simple messaging app in China, a cheerful way to send notes to one other, now helps its one billion users run much of their lives. As one of Andreessen's partners notes, they can use WeChat to "hail a taxi, order a food delivery, buy movie tickets, play casual games, check in for a flight, send money to friends, access fitness tracker data, book a doctor appointment, get banking statements, pay the water bill, find geo-targeted coupons, recognize music, search for a book at the local library, meet strangers . . . follow celebrity news, read magazine articles, and even donate to charity."[6]

But again, we should remember that the technology companies that populate the future might not be today's most familiar ones. Dominance today does not imply dominance in years to come. Back in 1995, for example, it was unthinkable that Microsoft's technological rule would ever come to an end, yet now they are being talked about as the "underdog" in the sector.[7] Nor do striking contemporary achievements necessarily mean that further remarkable successes will follow. For one cautionary tale, consider IBM's Watson, the celebrated *Jeopardy!*-winning computer system. Over the last few years, there has been great excitement about its broad potential. But despite their best efforts, a recent high-profile partnership between the Watson team and MD Anderson, a large American cancer hospital, ended in conspicuous failure: the $60 million system designed to help treat cancer was deemed "not ready for human investigational or clinical use."[8]

Indeed, the companies behind the health care technologies that really change our lives may not exist yet. And the same goes for the rest of the economy. After all, many of today's most familiar technology names—Airbnb, Snapchat, Spotify, Kickstarter, Pinterest, Square, Android, Uber, WhatsApp—did not exist a dozen years ago.[9] Many technologies that will be household names in the future probably have not yet been invented.

WHY BIG?

Like today's tech giants, the technology companies that dominate in the future are also likely to be very big. In part, this is simply because it costs an enormous amount to develop many of the new technologies. The best machines will require three expensive things: huge amounts of data, world-leading software, and extraordinarily powerful hardware. Only the largest companies will be able to afford all of these at the same time.

The first thing they will need is immense collections of data. We have seen examples of this at work already: AlphaGo, the first version of Google's go-playing system, learned in part from an archive of thirty million past moves by the best human players; Stanford's system for detecting skin cancer used almost 130,000 images of lesions, more than a human doctor could expect to review in their lifetime.[10] Sometimes, though, the necessary data is not readily available and has to be gathered or generated in costly ways. Consider what it takes to train and evaluate a car-driving system, for example. To do this, Uber built an entire mock town on the site of an old steel mill in Pennsylvania, complete with plastic pedestrians that occasionally throw themselves into traffic, and gathers data as their cars drive around it. Tesla, meanwhile, collects data from its *non*-autonomous cars as they are driven by their owners, with about a million miles' worth of data reportedly flowing in every hour. Google has adopted yet another approach to the problem, creating entire virtual worlds to gather data from cars driving around in these simulations.[11]

Then there is the matter of the software. Beneath all of these new technologies is the code that makes them work. Google's assorted Internet services, for instance, require two billion lines of code: if these were to be printed out on paper and stacked up, the tower would be about 2.2 miles high.[12] Writing good code requires talented—and expensive—software engineers. The average salary for a developer in San Francisco, for example, is about $120,000 a year, while the best engineers are treated as superstars and receive pay packages to match.[13] Today, when we recount economic history, we punctuate it with people like James Hargreaves, the inventor of the spinning jenny. In the future, when people tell the history of our own time, it will be filled with names like Demis Hassabis, of DeepMind, and other software engineers, as yet unknown.

As for processing power, many of the new systems require extraordinarily powerful hardware to run effectively. Often, we take for granted quite how demanding even the most basic digital actions we carry out can be. A single Google search, for instance, requires as much processing power as the entire Apollo space program that put Neil Armstrong and eleven other astronauts on the moon—not simply the processing power used during the flights themselves, but all that was used during planning and execution for the seventeen launches across eleven years.[14] Today's cutting-edge technologies use far more power still.

To be sure, some trade-offs can be made among the three requirements. Better software, for instance, can help compensate for a lack of data or processing power. AlphaGo Zero needed neither the data nor the processing power of its older cousin, AlphaGo, to beat it emphatically in a series of go games, 100–0.[15] How did it do this? By using more sophisticated software, drawing on advances in a field of algorithm design known as reinforcement learning.[16] The most powerful machines of the future, though, will probably be the ones that can draw on the best of all three resources: data, software, and hardware. And while small institutions may have one of them—perhaps a talented engineer capable of writing good software or a unique set of valuable data—they are unlikely to have all three at the same time. Only Big Tech will.

Aside from this issue of expensive resources, the dominant tech companies are also likely to be "big" for another reason: many new technologies benefit from very strong "network effects." This means that the more people who use a given system, the more valuable it becomes for those users. The classic explanation of these effects goes back to the days when telephones got installed in people's houses: adding a new person to a phone network is not only useful for her, but also valuable for everyone already on the network, because now they can call her as well. It follows that as a network grows, each additional person is even more valuable to the network than the one before. Mathematically, this idea is sometimes referred to as Metcalfe's law: the value of a network with n users is proportional to n^2.

Today, of course, we have moved beyond telephone landlines, and the obvious place to start when thinking about networks is with social media platforms. Facebook and Twitter, for instance, would be far less fun for their users (and far less lucrative for their owners) if there were no other people online to read what they share. This is also true of many other systems. Platforms like Airbnb and Uber become more valuable the more people there are using them: more apartments to rent and travelers looking for a place to stay, more cars to hire and passengers wanting a ride. What's more, they are built upon rating systems so that users can avoid a dud service—and, again, the more feedback there is, the more reliable such systems become. Think of the suspicion you might have about a solitary five-star rating with effusive praise on an arbitrary taxi website, versus the thousands of driver ratings on a platform like Uber.

Populous networks also let companies gather data that improves their products. Navigation systems like Waze and Google Maps gain a sense of the traffic on the road from the speed at which their users' phones move as they drive along. Amazon and Spotify tailor their purchasing and music recommendations based on what they glean from other people in their network like you. And then there is the basic bandwagon effect: once a particular network is popular, it makes sense to join it rather than a fledgling rival. I have a friend named Faiz who once thought about starting a new social network—but it is a curious person who would choose Faizbook, with just a few other patrons, over Facebook with its two billion users. Network effects do not make social platforms completely invulnerable—think of Friendster, MySpace, and BlackBerry Messenger, once-popular networks now consigned to the technological graveyard—but they certainly make it harder for small upstarts to gain traction.

All of this explains why large technology companies are acquiring so many other technology companies and start-ups. In the decade to July 2017, the Big Five made 436 acquisitions, worth about $131 billion.[17] These companies are trying to gather up valuable resources—particularly useful data, engineering talent, and network popularity—for themselves by buying them up when they see them on display in other companies.

THE ECONOMIC CASE AGAINST BIG TECH

For all the reasons described above, in the future our economies are likely to become dominated by large technology companies. Traditionally, the state has not relished the rise of this sort of economic dominance. It has developed competition policy, driven, in broadest terms, by the idea that monopolies are bad and competition is good.[18] Today's top technology companies already find themselves clashing with the authorities tasked with putting this policy into practice—because all of them aspire to monopoly power, if they have not obtained it by now.

This ambition is not unique to the world of technology. Look through the literature on management and strategy, and you will find plenty of ideas for achieving economic supremacy, packaged in the disarmingly benign-sounding language of business writing. Take Michael Porter, the definitive business strategy guru of the last few decades, whose 1980s books *Competitive Strategy* and *Competitive Advantage* were on the shelves of all discerning corporate leaders. Those books guided readers toward nothing less than economic domination: first, find markets ripe for monopolizing (or create new ones); second, dominate and exclude others from these chosen markets. Today, the same advice is given even more forthrightly. "Competition is for losers," wrote Peter Thiel, the entrepreneur, in the *Wall Street Journal*. "If you want to create and capture lasting value, look to build a monopoly."[19]

What, then, is the problem with an absence of competition? The economic argument advanced by competition authorities is that a monopoly in a given market means that our welfare is lower, both today and tomorrow. It is lower today because companies without competitors are able to inflate their profits, either by charging higher prices or providing poorer-quality products and services to customers. It is lower in the future because without being prodded by competition, these incumbent companies might be less willing to invest and innovate for the years ahead. These sorts of arguments are behind a handful of successful legal actions regarding anticompetitive behavior by Microsoft, Facebook, Apple, and Google, and there is speculation that Amazon is heading toward legal trouble as well.[20]

In practice, though, this economic argument in favor of competition is difficult to apply. First, it is not entirely clear what we mean by the word *welfare*. Do we only mean that consumers are happy or satisfied? If so, how do we measure that? The textbooks tell us to look at prices and imagine how much lower they might have been with more competition, but many of the large tech companies already give their products away for free. Secondly, it is often unclear what *market* we are actually talking about. Take Google, for example: if we think of it as being in the search engine business, then the fact that it controls 62.6 percent of search traffic and 88 percent of search advertising might set alarm bells pealing. But is this really Google's primary market? Given that Google gets most of its revenue from advertising, perhaps it is more accurate to think of it being in the advertising business. Then the competitive situation looks less troubling. The US search engine advertising market is worth about $17 billion a year, whereas the total US advertising market is worth $150 billion a year. Even if Google ended up owning the entire market for search engine advertising in the country, it would still have less than 12 percent of the American advertising business. "From this angle," writes Peter Thiel, "Google looks like a small player in a competitive world."[21]

In short, finding answers to even the most basic questions of competition policy is not straightforward. And perhaps the biggest complication of all is that monopolies can be a very *good* thing. This may sound like economic sacrilege, but the early-twentieth-century economist Joseph Schumpeter famously made just this case.

For Schumpeter, economics was all about innovation. He called it the "outstanding fact in the economic history of capitalist society." His argument for monopolies is that, were it not for the prospect of handsome profits in the future, no entrepreneur would bother to innovate in the first place. Developing a successful new product comes at a serious cost, in both effort and expense, and the possibility of securing monopoly power is the main motivator for trying at all. It acts as the "baits that lure capital on to untried trails."[22] Moreover, monopoly profits are not simply a consequence of innovation, but a means of funding further innovation. Substantial research and development very often draws on the deep pockets established by a company's past commercial successes. Think of Google, and its history of expensive failed ventures: Google Glass and Google Plus, Google Wave and Google Video. Just one of these flops would have broken a smaller company. But Google was able to withstand them, stay afloat, keep innovating and profit from the ventures that did end up succeeding.

Schumpeter was not troubled by concerns that monopolies might entrench themselves and lower welfare. Economists who worry about "nothing but high prices and restrictions of output" are missing the bigger picture, he said: economic dominance by any company is not a permanent state of affairs. In time, today's monopolies will be blown away by the "perennial gale of creative destruction." New ones will inevitably take their place, but only temporarily, for they, too, would eventually be broken up by the same storms.[23] These are the intellectual origins of the idea of "disruptive innovation," so popular today among management theorists and strategy consultants.

And Schumpeter was right: time and again, companies that seemed to be permanent fixtures of economic life have faded away. Take the

Fortune 500, an annual list of the five hundred largest corporations in the United States, which together make up about two-thirds of the American economy. If you compare the lists for 1955 and 2017, only about 12 percent of the companies made it from the first to the second. The other 88 percent went bankrupt, dissolved into other businesses, or collapsed in value and fell off the list.[24] Today, the names of corporations that did not survive—Armstrong Rubber, Hines Lumber, Riegel Textile, and the like—are unrecognizable, indistinguishable from the sort of fictional companies that fill the pages of novels. In their day, no doubt, they must have seemed like immovable giants. (This, again, is why we should not focus too much on the current Big Five tech corporations. Our attention should be on the more general problem, the fact that at any given moment in time a small number of tech companies are likely to dominate.)

Competition authorities are tasked with the balancing act of evaluating these various arguments for and against monopolies, judging the merits and dangers in any particular case of economic dominance. In the future, the nature of that task is likely to change dramatically. For instance, a few decades ago, if some companies wanted to collude and simultaneously raise their prices, it meant secret meetings and clandestine communications to coordinate their plans. Now, though, algorithms can monitor and move prices automatically, facilitating collusion without relying on old-fashioned anticompetitive tradecraft.[25] Indeed, today this may even happen unintentionally: a recent study found that the sorts of algorithms used by online firms to price their products might learn to implicitly cooperate with each other, keeping prices artificially high without any direct communication and without any instructions to collude.[26] Whether such algorithmic behavior should be targeted by competition policy is an open question.

Likewise, consider how, in the past, competition authorities might have viewed a prolonged period of outsize profits to be a sign that a large company was abusing their economic clout. Yet today, some companies seeking economic power intentionally endure long periods of staggering *unprofitability*—pursuing rapid growth and competition-neutering expansion, attempting to crowd out their competitors through scale and dominance. As Figure 11.1 shows, for instance, for most of its history Amazon made almost no profit.[27] Uber has followed in their footsteps,

Figure 11.1: Amazon's Annual Revenue and Net Profit, 1998–2018 ($bn's)[28]

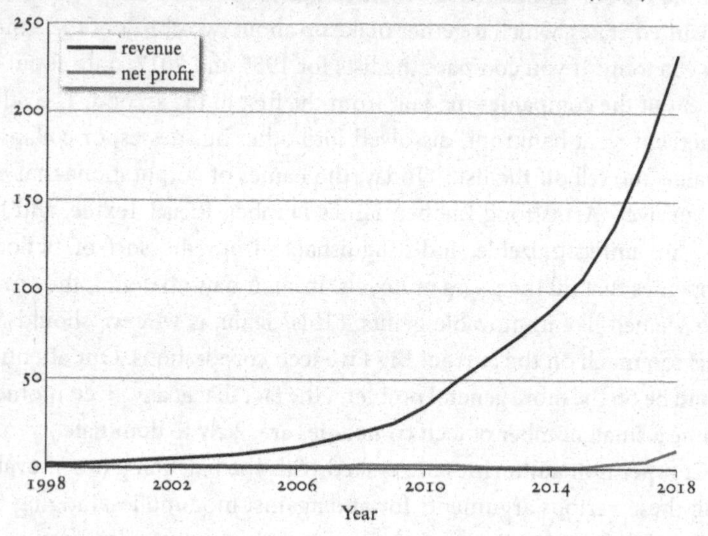

failing to make a profit each year since it was founded.[29] Authorities should not be thrown off the scent: at this point, traditional economic indicators like the level of profit may no longer be a reliable guide to anticompetitive behavior.[30]

As large technology companies continue to become more dominant through the twenty-first century, it is inevitable that they will clash more frequently, and intensely, with competition authorities. Some of these firms, no doubt, will be in breach of competition and antitrust laws, will become too economically dominant, and will have to be broken up accordingly. That said, in decades to come the most compelling case against Big Tech will not be this economic one. Rather, as technological progress continues, our concern will shift from the *economic* power that these companies wield—however mighty it may be—to their *political* power instead.

THE POLITICAL CASE AGAINST BIG TECH

Commentators today often like to draw a comparison between Big Tech and Standard Oil, the American giant founded by John Rockefeller in 1870. When that company was established, it owned the largest oil refinery in the world.[31] By 1882, it controlled 90 percent of the country's oil production.[32] Its dominance continued until 1911, when the United States Supreme Court stepped in on antitrust grounds. In perhaps the best-known case of its kind, it concluded that Standard Oil had become a monopoly, and broke it up into thirty-four smaller companies.

It is easy to see the appeal of the analogy. Standard Oil's dominance of oil refining in the United States at the turn of the twentieth century is similar in scale to the way that today's Big Five tech firms dominate their respective industries. And there is a certain similarity to their operations, too: companies like Facebook and Google control the flow of valuable personal data, seen as "the oil of the digital era."[33]

Yet this comparison is most revealing for what it *fails* to capture. Think about the nature of the legal objections to Standard Oil's dominance: at the time, they were overwhelmingly economic ones. In the language of the Supreme Court, the accusation was that Rockefeller's company was engaged in "an unreasonable or undue restraint of trade." They had gained too much economic power and were using it to distort the oil market.[34] By contrast, when we examine the sort of objections that are increasingly leveled at Big Tech, unlike Standard Oil, they often have very little to do with economics at all.

Consider some concerns that have been aired about Google. If you search for an African American–sounding name, for instance, you are more likely to be served up an advertisement for criminal background checks.[35] A few years ago, if you tried to use Google's image recognition algorithm to tag photos, it might have labeled Black people as "gorillas"; the company has dealt with the problem just by removing the tag "gorillas" from its algorithm altogether.[36] More generally, there are worries about Google's ability to promote and demote particular websites returned by a search; reportedly, this ability has been used to remove file-sharing sites from search results.[37] In the last few years YouTube, owned by Google, has been criticized for recommending far-right videos and anti-vaccine channels, tolerating hate speech and encouraging pedophiles.[38]

Or take Facebook. The company ran an internal experiment that showed that displaying positive or negative stories could influence the emotions of its users; the 689,000 users involved did not know they were lab rats.[39] By looking at a person's "likes" alone, it is possible to guess their sexuality correctly 88 percent of the time; this technique is said to be useful for advertisers to tailor what they show and to whom they show it.[40] In fact, Facebook was sued by the US Department of Housing and Urban Development for allowing advertisers to intentionally target ads by race, gender, and religion; these groups still receive different ads from one another.[41] During the 2016 US presidential election, Russia bought Facebook ads and set up groups to fire up political division among voters; Facebook identified several thousand of them, but only after the damage was done.[42] In a study of over three thousand anti-refugee attacks in Germany, researchers showed that regions with higher Facebook usage experienced significantly more attacks, with online hate speech from the far-right Alternative für Deutschland (AfD) party translating into real-life violent crimes.[43]

Then there is Amazon. In 2009, after a dispute with an electronic publisher, they logged into every Kindle device and deleted any e-books users had bought from that company; ironically, among these was George Orwell's *1984*.[44] In 2017, Amazon put on sale a set of iPhone covers with images such as heroin-filled needles, an old man with a crutch wearing a giant diaper, and a close-up of five toenails infected with fungus; an algorithm was pulling stock images from the Internet and creating products without any human supervision.[45] In 2015, an Amazon Echo device—which records any commands or questions put to it—was the sole "witness" to an alleged murder in Arkansas, and the prosecution sought to obtain the recording of that night's interactions. (The case was eventually dismissed.)[46]

Or take Apple. The company has complete control over what apps can appear on the iPhone: they refused to host an app that is critical of their manufacturing methods, but supported an app that is highly critical of climate change science; they banned an app that helps its users track US drone strikes, but allowed an app that helps men in Saudi Arabia track women and limit their movements.[47] In 2016, Apple refused to help the US government to unlock the iPhone of one of the terrorists involved in the San Bernardino mass shooting, arguing that forcing them to write new software to break the encryption was a violation of their freedom of speech.[48]

And, finally, think about Microsoft. In 2016, they launched a Twitter chatbot called Tay, which was designed to learn from other Twitter users and imitate the way a teenage girl might speak. It was very swiftly removed after it began to post racist slurs, praised Hitler, denied the Holocaust, and made sexually explicit comments to its followers. Microsoft said it was "making some adjustments."[49] A year later, it released a new chatbot, called Zo, which was programmed to avoid talking about politics and religion. Yet Zo quickly showed the same troubling tendencies: "the quaran [*sic*] is very violent," it wrote in response to a journalist's question about health care. Microsoft kept Zo active for two and a half years.[50]

Some of these stories may seem abhorrent. Others might appear less troubling, or perhaps even amusing. But what all of them share is that they have very little to do with economic power, and with the measures of consumer welfare that might keep an economist up at night. Instead, they are worries about the role that these new technologies might have in distorting the social structures that support our shared existence. They are, in short, concerns about *political* power. They are worries that these large technology companies, and not the society in which they operate, are in control of shaping how we live together.

In the case of Standard Oil, political power was not really the concern. Anxieties focused on their economic power, on the possibility that the oil they sold would become too expensive because there was not enough competition in the market. Yes, some critics of Standard Oil accused the company of having a malign influence on US politics. But there is something fundamentally different about Big Tech's political power today, as Jamie Susskind—a lawyer and political theorist who also happens to be my brother—shows in his book *Future Politics.*

The word *politics* is sometimes used in a narrow way, denoting just the hustle of politicians and the decision making of the state. That sense is what critics of Standard Oil's political power had in mind. But politics, properly understood, is much bigger than that. It is about how we live together in society, about all the different forces—not only the people and institutions of the traditional political process—that shape our collective lives. This is why when we call something *political* we tend to also mean that it is very important. The women's liberation movement of the 1970s, for instance, understood this very well.[51] They fought to make the world wake up and see that the personal part of our lives—sex and relationships, child care and housework, fashion and pastimes—truly matters, that "the personal is political." And in that spirit, Jamie Susskind writes that these days "the digital is political."

In the future, Big Tech will be ever more politically powerful in this broader sense. As described in *Future Politics*, these companies will set the limits of *liberty*—think of a driverless car that cannot go above a certain speed, for example. They will shape the future of *democracy*—think of an electorate reared on political facts curated to their personal tastes and served up by algorithms. And they will determine questions of *social justice*—think of someone who finds their request for a financial loan or health treatment turned down on the basis of personal data they never agreed to give away.[52]

In the twentieth century, our main preoccupation was with the economic power of large companies. But in the twenty-first, we will increasingly have to worry about this political power as well. New technologies may begin in the marketplace, with products that people are happy to pay for and put to personal use. But their effects will seep out and shape our shared lives as political creatures, too. In the past, questions of liberty, democracy, and social justice were answered by us, as citizens, and by our political representatives, living together in civil society. In the future, unless we act, those decisions will increasingly be made by engineers, doing their work out of sight at large technology companies. The threat, in short, is the "privatization" of our political lives.[53]

THE POLITICAL POWER OVERSIGHT AUTHORITY

Competition policy might involve ambiguous concepts that are cumbersome to apply, but it still provides a rough framework for action. We have an intuition about what too much concentrated economic power looks like, and we have sensible, well-rehearsed ideas about what to do in response. In the event of disagreement, at least we know what it is we are quarreling about. At the moment, though, nothing comparable exists to help us think clearly about political power. The instinctive sense of unease most of us feel when we reflect on some of the examples above shows that we know something troubling has begun. But we do not know precisely how misuse of political power should be defined, and we do not have any systematic ways of responding.

What parts of our political lives should be shaped by these new technologies, and on what terms? The problem is that, at the moment, we leave it almost entirely to Big Tech to answer. While we impose tight constraints on the economic power of these companies, they are free to choose much of their own noneconomic behavior as they move onto this new political turf. We let them both set their own boundaries and police those boundaries. Executives from these companies increasingly sit on the commissions, boards, and bodies tasked with exploring the wider consequences of these new technologies, and understandably appear to be quite content to handle the problem themselves. Google's CEO, Sundar Pichai, for instance, accepts that fears about AI are "very legitimate," but believes companies like his should "self-regulate."[54] This attitude is widespread.

But do we really trust Big Tech to restrain themselves, to not take advantage of the political power that accompanies their economic success? And even if they wanted to act to constrain their political power, are they actually *capable* of doing it? These companies may have the deep technical expertise needed to build these new systems, but that is a very different capability from the moral sensibility required to reflect on the political problems that they create. Software engineers, after all, are not hired for the clarity and sophistication of their ethical reasoning.

There are some on the political left who say that rather than leaving such decisions to Big Tech, we should nationalize the companies, giving control of entities like Google and Facebook over to the state.[55] But that proposal neglects something very important (among many other problems with the idea): there is little reason to think that the state would be immune from similarly abusing the political power created by these new technologies. Consider China's rollout of a "social credit system," for example: the government's ambition is that by 2020, all Chinese citizens will be scored and ranked according to information about them stored in a national database. In one pilot, scores are determined by acts as trivial as getting a traffic ticket (minus 5 points) and as ambiguous as "committing a heroic act" (plus 30 points). Images of "civilized families" with high social credit scores are posted on public notice boards.[56] Officials say that they want the system to "allow the trustworthy to roam everywhere under heaven while making it hard for the discredited to take a single step."

Or, for a more mundane example, take the issue of data security. Today, headlines are dominated by stories of Big Tech mishandling our personal data. But not so long ago, the national conversation in the UK was preoccupied with cases of public officials losing our personal data instead. At one point, a UK official was being fired or sanctioned on average *every working day* for mishandling sensitive data.[57]

Instead of nationalization, then, what is needed is a new regulatory institution—set up in the spirit of competition authorities that regulate the economic power of these large companies, but tasked with constraining their political power instead. Think of this as the Political Power Oversight Authority.

The first task for this new agency would be to develop a framework that allows regulators to identify, in a clear and systematic way, when political power is being misused. Competition policy does this for economic power; we need something analogous in this new political sphere. When is a restriction of our liberty so egregious, a threat to our democratic process so severe, or an instance of social injustice so utterly unacceptable that an intervention is required? These are big questions. The ambiguities that competition authorities face in answering their question—"are consumers better off or not?"—might seem pedestrian in comparison. But complexity is not an excuse for inaction; these new questions demand a response. At the same time, though, the new authority must not lurch into overreaction. The goal of the Political Power Oversight Authority should not be to strip Big Tech of any political power altogether. Just as competition authorities take into account both the merits and dangers of economic power, so, too, this new authority must perform a similar balancing act. These new technologies, after all, are responsible for improving our lives in countless ways as well.

But aren't people happy to use Big Tech's offerings and services? And doesn't that mean they have consented to all of the political consequences of those technologies as well? No. As *Future Politics* makes clear, the key question is whether Big Tech's political power is legitimate—and the fact that people are happy to use Big Tech's products and services is not enough to establish consent. Consumer satisfaction might justify the companies' economic power, their profits and executive pay packages, but political power should not be bought or sold in this way. The fact that people enjoy posting on Facebook does not authorize Facebook to ignore the way its platform gets used for sinister political ends. The fact that people like using Google's search engine does not mean that Google can turn a blind eye when their ads discriminate against users. Economic success is not a free pass to run roughshod over our political lives.

The Political Power Oversight Authority must have a spread of capabilities at its disposal if it needs to act. It will need investigative tools; the ability to inspect particular companies and scrutinize their technologies to determine whether political power is excessive or being abused. Then there must be transparency tools, to compel companies to be open about their operations and offerings: it is not possible for people to properly consent to new technologies if they do not know, for instance, what data about them is collected and shared, how it gets used, or even which vendors are developing those systems. More powerful tools would allow the new agency to mandate or restrict certain types of behavior; and the most powerful would permit it to call for the breakup of big companies if their political power is judged to be too great. None of these capabilities are particularly revolutionary: competition authorities use very similar tools to police economic power today. The task now is to empower regulators to apply a version of them in this new political arena as well.

Importantly, this new authority must be distinct from our traditional competition authorities. This problem is a political one, not an economic one, and the economists who tend to populate our existing agencies are not the right people to grapple with this challenge. The conceptual tools they deploy to think about prices and profits, however insightful and effective they may be, tell us nothing at all about ideas like liberty, democracy, and social justice, and whether they are under threat.

As an economist myself, I might seem to be shooting myself in the foot here. But listening to some economists today attempt to discuss these political problems—whether by arguing that these are actually economic problems after all, or by claiming to have expertise in political issues as well—can be a painful experience. We all have to recognize that solving these new challenges will require very different people from those who were best equipped to handle the challenge of economic power in the past. We need a new institution, staffed by political theorists and moral philosophers, to watch over individuals as *citizens* in a society, not simply as *consumers* in a marketplace. That is what this new authority must do.[58]

12

Meaning and Purpose

There is an old joke about an elderly Jewish mother who finds herself at the seaside with her adult son. He heads into the sea for a dip, but it turns out that he is a bad swimmer. Drifting farther away from the shore, he begins to panic and struggle in the water. And his mother, watching the trouble unfold from the beach, turns around and shouts out to everyone around her: "Help, my son, *the doctor*, is drowning!"

So far in this book, there has been no room for this mother's pride in her son's profession. I have looked at work from a purely economic perspective, where it only matters because it provides an income. This perspective is helpful because it makes the threat of technological unemployment very clear: by doing away with work, automation will deprive people of their livelihoods. But for some, like the anxious mom at the seaside, this will seem like a shallow account of why work is important. They will feel that the issue goes beyond economics, that a job is not simply a source of income but of meaning, purpose, and direction in life as well.

From this viewpoint, the threat of technological unemployment has another face to it. It will deprive people not only of income, but also of significance; it will hollow out not just the labor market, but also the sense of purpose in many people's lives.[1] In a world with less work, we will face a problem that has little to do with economics at all: how to find meaning in life when a major source of it disappears.

MEANINGFUL WORK

In fairness to economists, not all of them have always used such a ruthlessly narrow conception of work. It is true that in economics textbooks today, work is treated as an inescapably unpleasant activity that people do purely for the sake of an income; it causes "disutility," or unhappiness, that is only offset by the happy fact it earns a wage in return. And this sort of view has a long intellectual history, stretching back to Adam Smith, who once described work as "toil and trouble."[2] But others have thought differently. Take Alfred Marshall, another giant of economic history. He proclaimed that "man rapidly degenerates unless he has some hard work to do, some difficulties to overcome," and that "some strenuous exertion is necessary for physical and moral health." To him, work was not simply about an income, but the way to achieve "the fullness of life."[3]

And beyond economics, some of the great scholars have written at length about work and meaning. Sigmund Freud is often credited with saying that human well-being only depends on two things, "love and work."[4] What he actually wrote, though, was characteristically more esoteric than that—"the communal life of human beings had, therefore, a two-fold foundation: the compulsion to work, which was created by external necessity, and the power of love." Freud believed that work is "indispensable" for human beings, not so much for an income as for letting us live harmoniously in society: it is a necessary outlet for the deeper, primal impulses that everyone carries within them. Better to punch away violently on a keyboard in an office cubicle, Freud seems to suggest, than at each other's faces.[5]

Another figure famed for his reflections on work and meaning is Max Weber, the classical sociologist. Why do people attach so much meaning to the work that they do? Because of religion, Weber said—in particular, because of the sixteenth-century Protestant Reformation. Before then, Christians in Western Europe were mostly Catholic. If they felt guilty about what they had done (or thought), they could fix it through confession: sit down with a priest, share your sins, and the Church would absolve you of your wrongdoings and rescue you from damnation. For Protestants, though, this was not an option. They did not do confession. And this led to a "tremendous tension," Weber proposes, since people never knew whether they would be damned to burn in hell for eternity.[6] For them, the only relief was "tireless, continuous and systematic work," through which they could try to prove that their souls were worth saving.[7] Weber spoke of work as a "vocation" and a "calling," a task "given by God," all terms that are still used today.[8] In his view, the commitment to their work that some people show is, quite literally, a form of religious devotion.

Perhaps the most intriguing empirical examination of work and meaning is a study carried out by Marie Jahoda, a social psychologist, in the 1930s.[9] Its setting was Marienthal, a small village outside Vienna, founded in 1830 to provide homes for workers employed in a newly built flax mill nearby. As the factory grew over the following decades, so did the village. But in 1929, the Great Depression hit. The next year, the factory closed down. By 1932, three-quarters of the 478 families in the village had nobody in work, and relied entirely on unemployment payments for an income.

Jahoda and her colleagues wanted to know what the impact of such widespread worklessness would be. Their methods were unconventional: to collect data on residents without making them realize they were being watched, the researchers embedded themselves in everyday village life. (Their various enterprises included a clothes cleaning and repair service, parent support classes, a free medical clinic, and courses in pattern design and gymnastics.) What they found was striking: growing apathy, a loss of direction in life, and increasing ill will to others. People borrowed fewer library books: 3.23 books on average per resident in 1929, but only 1.6 in 1931. They dropped out of political parties and stopped turning up to cultural events: in only a few years, the athletic club saw membership fall by 52 percent and the glee club by 62 percent. Unemployment benefits required that claimants do no informal work; in those years, Marienthal saw a threefold increase in anonymous denunciations of others for breaking that rule, yet almost no change at all in the total number of complaints that were judged well-founded.

Researchers watching at a street corner even noted a physical change: men without work walked more slowly in the street and stopped more frequently.

For Freud, then, work was a source of social order; for Weber, it provided people with a greater purpose; for Jahoda, it created a sense of structure and direction. And alongside these scholarly reflections we can also put more familiar, everyday examples of people searching for purpose in the work that they do. Walk into a bookstore, and you will find countless books telling readers how to achieve fulfillment in their working lives. Apply for a new job, and an eager employer may tempt you by promising not simply a healthy income but a meaningful career. Chat with a proud breadwinner, and they will mention the glow of earning one's keep or supporting a family. Talk to new parents about how it feels to leave their job for a new role at home, even temporarily, and they will often refer to a sense of loss that goes well beyond the value of the wages they once received. Look at prosperous people who could afford never to work again, and you will see many still get out of bed and go into the office on a daily basis, often after a brief and unsuccessful experiment with retirement. Sit down at a dinner party, and a stranger may ask, "What do you do for a living?"—often presuming that the work that you do says something significant about who you are.

This last observation is important. Work matters not just for a worker's own sense of meaning; it has an important social dimension as well, allowing people to show others that they live a purposeful life, and offering them a chance to gain status and social esteem. Today, social media has supercharged this phenomenon. It is no wonder that LinkedIn, which started as a networking tool to help people find new jobs, is now used by some to broadcast how successful they are and how hard they are working, turning into a platform for a kind of conspicuous self-aggrandizement instead.

For those with a job, the connection between work and meaning is wonderful: in return for their efforts, they get both an income and a sense of purpose. But for the unemployed, this link may become instead a source of further discomfort and distress. If work offers a path toward a meaningful life, the jobless may feel that their existence is meaningless; if work provides status and social esteem, they may feel out of place and deflated. This may partly explain why the unemployed often feel depressed and shamed, and why their suicide rate is about two and a half times the rate of those in work.[10]

A prevailing political philosophy of our time, the idea of meritocracy, does little to help.[11] This is the notion that work goes to those who somehow deserve it, due to their talents or effort. Yet if work signifies merit, then those without it might feel merit*less*. Michael Sandel once quipped that in feudal times, at least those at the top knew that their economic fortunes were a fluke of birth, the simple brute luck of being born into the right family—whereas today, the most fortunate imagine they actually merit their positions, that being born with the right talents and abilities (and, often, supportive and prosperous parents) has nothing to do with luck at all.[12] An unpleasant corollary is that the less fortunate now often think they merit their bad luck as well.

At times, work does seem to get its meaning not from the positive thought that having a job is something to celebrate, but from the opposite, negative idea—that to be without a job is worthy of shame. When people call the unemployed "benefit scroungers" or "welfare queens," they are of course stigmatizing those without work, but they are also reinforcing the esteem given to those who are in work. And while it might feel that resentment of the unemployed is a new phenomenon, whipped up by tabloid newspapers in the twentieth century, it is in fact a very old sentiment. It is evident, for instance, in the Poor Laws, a body of rules that started to coalesce in medieval Britain, and introduced the first government taxes earmarked for helping the poor. (Before then, support for the poor was largely voluntary and informal, provided by friends, family, and the Church.) An early edition of the laws, from 1552, stated, rather dramatically, that "if any man or woman, able to work, should refuse to labor and live idly for three days, he or she should be branded with a red hot iron on the breast with the letter V and should be judged the slave for two years of any person who should inform against such idler."[13]

The resentment runs both ways. While those in work rail against the unemployed, those without work also feel aggrieved toward those with it. This, in part, explains the curious reaction to Silicon Valley's recent enthusiasm about the UBI. Mark Zuckerberg and Elon Musk have made supportive noises about the idea of a UBI; Pierre Omidyar, founder of eBay, and Sam Altman, founder of Y Combinator, have funded trials of it in Kenya and the United States.[14] But their interest has been met with widespread hostility. If work were simply a means to an income, that response might seem odd: these entrepreneurs were essentially proposing that people like them should do all the hard work and give everyone else money for free. For many people, though, work means more than securing a wage—and so, in their eyes, the offer of a UBI from those in fantastically well-paid jobs might have felt more like hush money, or a bribe, perhaps even an attempt to monopolize a source of life's meaning and deny it to others.

MEANINGLESS WORK

The connection between work and meaning is powerful indeed in many parts of today's working world. Yet it is not universal—and what's more, where it does exist it seems to be a relatively recent phenomenon.

Our prehistoric ancestors, for instance, might have found the idea that work and meaning are tied together very strange. Until the 1960s, it was thought that hunter-gatherers must have lived highly laborious lives, but recent anthropological research has shown that they probably did "surprisingly small" amounts of work. When Gregory Clark, an economic historian, reviewed a range of studies about contemporary hunter-gatherer societies, he found that their members spent consistently less time engaged in labor than today's average male worker in the UK. (Clark's definition of labor includes not just paid employment but study, housework, child care, personal care, shopping, and commuting.)[15] The data imply that hunter-gathers in subsistence settings tend to take about a thousand more hours of leisure a year, on average, than working men in UK's prosperous modern society.[16]

This is not what you would expect if the hunter-gatherers were relying on work to find purpose and fulfillment. Clearly, they simply sought—and continue to seek—life's meaning elsewhere. As anthropologist James Suzman puts it, "the evidence of hunting and gathering

Figure 12.1: Male Labor Hours per Day, Hunter-Gatherers and the UK Today[17]

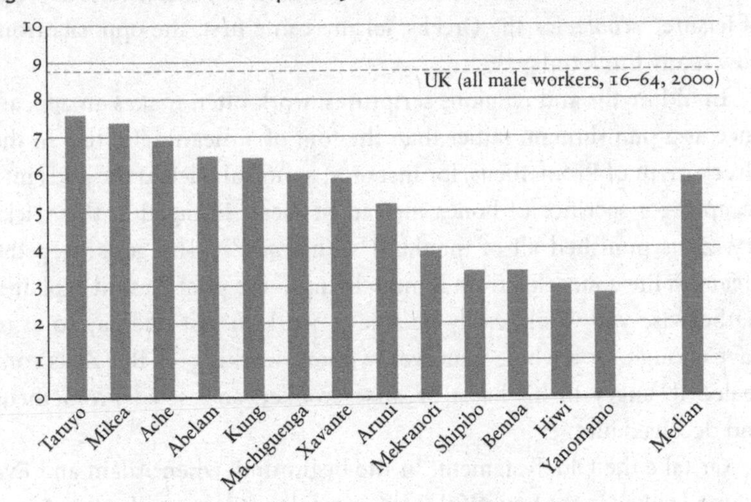

societies suggests . . . we [human beings] are more than capable of leading fulfilled lives that are not defined by our labor."[18]

The attitudes were different in the ancient world, too. Back then, work was often thought to be degrading rather than meaningful.[19] In the ancient Egyptian city of Thebes, the law stipulated that nobody could hold office unless he had kept away from trade for ten years.[20] Handling goods in the market was considered to be a prohibitively grubby affair. In Sparta, the warrior city-state of Greece, citizens were raised to fight and kept away from productive work by law. Trade was left to noncitizens, and manual labor to the "helots," a vast state-owned population of slaves.[21]

When Plato set out his blueprints for his ideal state, he confined certain workers to their own "artisan class," denying them a chance to run state affairs. "The best-ordered state will not make an artisan a citizen," he said. Aristotle, likewise, wrote that "citizens must not lead the life of artisans or tradesmen, for such a life is ignoble and inimical to excellence."[22] He believed that meaning could only come through leisure, and that the only purpose of work is to pay for leisure time: "We work in order to enjoy leisure, just as we make war in order to enjoy peace."[23]

In fact, the Greek word for "work," *ascholia*, literally means "the absence of leisure," *schole*; for the Greeks, leisure came first, the opposite from how many think today.[24]

In old myths and religious scriptures, work often makes an appearance as a punishment, rather than the font of a meaningful life. In the Greek myth of Prometheus, for instance, a mortal tricked the gods into accepting a sacrifice of bones instead of meat. Enraged at this trickery, Zeus punished all of mankind with *work*.[25] "The gods keep the means of life concealed from human beings," the poet Hesiod explains. "Otherwise you would easily be able to work in just one day so as to have enough for a whole year even without working . . . But Zeus concealed it, angry in his heart because crooked-counseled Prometheus had deceived him."[26]

Or take the Old Testament. In the beginning, when Adam and Eve roamed naked in the bountiful Garden of Eden, all was well. After Adam ate the forbidden apple, though, God condemned them both to hard labor—Eve, metaphorically, to hard labor through painful childbirth ("I will make your pains in childbearing very severe"), and Adam, literally, through making him toil from then on for his sustenance ("by the sweat of your brow you will eat your food").[27]

These stories should remind us that the link between work and meaning, no matter how much Freud and Weber may extol it, may not in fact be so clear. In blunter terms, for many people work has always been pretty miserable, whatever theory might suggest. It is hard to argue that toiling in the factories and mills of the Industrial Revolution gave people a deep sense of fulfillment, for example. On the contrary, it was a life of gloom and despair. This is what enraged a young Karl Marx, prompting him to write at great length about "alienation," the idea that certain work stopped human beings from becoming their true selves.[28] This is why Adam Smith, so often held up as the standard-bearer for unfettered markets, nonetheless feared that the dull monotony of labor would cause people to turn "as stupid and ignorant as it is possible for a human creature to become."[29] And this is why Charles Fourier, an influential early-nineteenth-century French philosopher, described the working world of the time as a "veritable graveyard."[30]

Nor do we have to go back to the Industrial Revolution, when workers lacked legal protections and were exploited and oppressed, to question the relationship between work and meaning.[31] Look at how people spend their working lives today, whether it's stacking shelves or making sandwiches, sweeping roads or collecting garbage, drafting legal contracts or reviewing financial accounts. Though these endeavors may not resemble the activities on the factory floor of a hundred years ago, none of them are self-evidently roles that bring their bearers a profound sense of meaning or fulfillment. In the United States, almost 70 percent of workers are either "not engaged" in or "actively disengaged" from their work, while only 50 percent say they "get a sense of identity from their job."[32] In the UK, almost 40 percent of people think their work does not make a meaningful contribution to the world.[33] In the words of the sociologist David Graeber, many people today find themselves trapped in "bullshit jobs."[34]

Finally, even for those who are fortunate and privileged enough to find their jobs meaningful, it does not follow that they would want to work if they did not have to. Take the French. They attach more importance to their work than many other nationalities. Yet, at the same time, they wish to—and do—devote less time to it than people in most other countries.[35] At times, I wonder whether the academics and commentators who write fearfully about a world with less work are just mistakenly projecting the personal enjoyment they take from their jobs onto the experience of everyone else.

THE OPIUM OF THE PEOPLE

There are, then, two very different views of the relationship between work and meaning. There are those who imagine that there is an important connection between the two, who see work as a way not only of distributing income in society but of sharing out meaning as well. Reading the downbeat accounts of "bullshit jobs" and miserable employment, they are likely to instinctively feel that things need not be that way, that with the proper changes even unpleasant work could be made fulfilling, too. Then there are those who have the reverse inclination, who question any link between work and meaning, and are likely to see any unhappiness or disillusion with work as confirmation of their beliefs.

In a sense, though, it may not matter which view you take. As we approach a world with less work, both tribes will be forced to address the same question: If people no longer work, what exactly will they do?

A popular idea for answering this question has been to look to the wealthy upper classes for guidance. Throughout history, many of their members have had the luxury of not having to worry about work for an income. Keynes called them "our advanced guard," imagining them up ahead in a world with less work, "spying out the promised land for the rest of us and pitching their camp there."[36] Wassily Leontief invoked them as well, writing that "those who ask what the average working man and woman could do with so much free time forget that in Victorian England the 'upper classes' did not seem to have been demoralized by their idleness. Some went hunting, others engaged in politics, and still others created some of the greatest poetry, literature, and science the world has known."[37]

Bertrand Russell, a philosopher and a member of the British upper classes himself, captured his views about his prosperous peers in a famous essay, "In Praise of Idleness." He argued that "a great deal of harm is being done in the modern world by belief in the virtuousness of work," and that "the road to happiness and prosperity lies in an organized diminution of work." It seemed to him that the leisure class "contributed nearly the whole of what we call civilisation . . . Without the leisure class, mankind would never have emerged from barbarism."[38] He thought no one should be obliged to work more than four hours a day, leaving people free to devote themselves to the arts, sciences, literature, and economics.

The problem is that looking at the lifestyles of the well-to-do is not particularly revealing. For one thing, there is a tendency to romanticize how wisely they actually spent (or spend) their days. Thorstein Veblen, whose theory of conspicuous consumption mocked how prosperous people in Victorian Britain spent their money ("in order to be reputable it must be wasteful"), also poked fun at how they occupied their free time—what he called "conspicuous leisure."[39] For them, it was not enough to be seen wasting income on garish superfluities; they had to be seen wasting their time as well. That explained the attachment of the "leisure classes" to everything from learning ancient languages to putting on elaborate exhibitions of punctilious manners and decorum. Of course, Veblen was being provocative. But he had a serious point, too: the upper classes often did spend their time in peculiar ways.

The truth is that imagining how we might spend our time profitably is very difficult. There is a famous line in Marx's writing that hints at why: "Religion," he says, "is the opium of the people." This is often interpreted as an attack on the clergy and the upper classes—blaming them for cooking up religious doctrine that sedated working people, blinding them to economic inequality and keeping them from starting a revolution. But this is not what Marx meant. He thought religion was created by ordinary people, not imposed top-down by others. It was their self-concocted way to add meaning to their lives.[40]

Today, though, religion clearly no longer fulfills that role. Religiosity may be on the rise in certain communities, and some new additions have been made to the religious canon (the "scripture" of Scientology, for example). But compared to Marx's age, a time of widespread church construction and enthusiastic clergy recruitment, the modern world is very different.[41] Religion no longer dominates everyday life in the way it once did.

What has taken its place? It is the *work* that we do. For most of us, work is the new opium. Like a drug, it provides some people with a pleasurable burst of purpose. But at the same time, it intoxicates and disorientates, distracting us from looking for meaning elsewhere. This makes it difficult to imagine how we could live our lives any differently. Work is so entrenched in our psyches, we have become so dependent on it, that there is often an instinctive resistance to contemplating a world with less of it, and an inability to articulate anything substantial when we actually do.

Most of us know what it means to live in gainful employment; the same cannot be said of what it means to be settled into gainful unemployment. In the words of Hannah Arendt, we live in a "society of laborers which is about to be liberated from the fetters of labor, and this society does no longer know of those other higher and more meaningful activities for the sake of which this freedom would deserve to be won." The worry, as Keynes put it, is that "there is no country and no people, I think, who can look forward to the age of leisure and of abundance without a dread. For we have been trained too long to strive and not to enjoy."[42]

LEISURE POLICIES

So, what should we do? How will people without work spend their time, and what can we do—if anything—to make sure that it provides them with the sense of purpose they might have hoped to find in a job? One of the most haunting findings from Marie Jahoda's time in the village of Marienthal was the way that leisure became what she called a "tragic gift" to those without work. Her hope had been that "even amid the misery of unemployment, men would still benefit from having unlimited free time." Instead, she found that, "cut off from their work . . . [they] drift gradually out of an ordered existence into one that is undisciplined and empty," to the point that when asked to explain how they spent their days, they were "unable to recall anything worth mentioning."[43] In a world with technological unemployment, how do we avoid a similar despondency and despair from spreading?

Part of the answer is that, as a society, we will need to think more carefully and consciously about leisure: both how to prepare for it, and how to use it wisely and well. Today, we are familiar with *labor market policies*, a spread of interventions that shape the world of work in a way that society thinks best. As we approach a world with less work, though, I believe we will want to complement them with something different: *leisure policies* that inform and shape the way that people use their spare time.

Revisiting Education

A serious leisure policy must begin with education. Today, the priority for most schools and universities is preparing people for the world of work. (And where that is not the aim, it is still how they are judged.) With that in mind, I argued earlier that we need to change what, how, and when we teach. But as we approach a world with less work, that particular priority will no longer make sense. Education experts are fond of quoting the ancient Spartan king Agesilaus, who said that the purpose of education is to teach children the skills they will use when they grow up.[44] Their point in citing the king's seemingly self-evident advice tends to be that today's education systems fail to do that. But when considering a world with less work, that same quotation prompts a different thought: the skills needed to flourish in that future world will be very different from those in demand today.

At the moment, we tend to conflate working and flourishing. We believe that to succeed in work is to flourish in life, so the skills required are the same. But if there is less work to be done, we need to prepare people for that instead. This would require a significant overhaul in what we teach—but big shifts of this magnitude are not unprecedented. If we traveled back to Agesilaus's time and took a look at the Spartan curriculum, we would be startled. Known as *agoge*, it was essentially a two-decade-long PE class designed to prepare men for war. Today, to the dismay of no one except perhaps PE teachers, organized exercise in schools is confined to a few hours a week. We no longer need to train young people to be warriors. In the future, we may no longer need to train them to work, either, but will have to teach them to flourish through leisure instead.

There are more recent sources of inspiration than the ancient Greeks, as well. About three-quarters of a century ago, the Education Act of 1944 introduced free secondary education for all in the UK.[45] The legislation's chief architect, an MP named Rab Butler, stood up in the Houses of Parliament to say that he hoped the reform would "develop our most abiding assets and richest resources—the character and competence of a great people." His wording suggests a dual aspiration: not simply to produce ever more competent workers, but also people of outstanding character.[46] In the decades that followed, the education system largely shed that latter aim, but in recent years it has made a comeback. Politicians, academics, and policy wonks increasingly talk about the importance of teaching "character" and "life skills" in school. One think tank, channeling the ethos of the classical philosophers, argues that we need to cultivate a set of "virtues" in students: "moral virtues such as honesty and kindness, civic virtues such as community service, intellectual virtues such as curiosity and creativity, and performance virtues such as diligence and perseverance."[47] We might argue whether these are the exact skills needed to flourish in a world with less work. But this basic exercise, revisiting the role of education beyond basic workplace competence, is the right one.

Shaping Leisure

Beside preparing children for a world with less work, societies may also want to devise leisure policies that shape how jobless adults actually spend their spare time. That might sound like a step too far: it is all very well for the state to try to influence the labor market, you might think, but should people not be left alone to choose how they spend their leisure? I am not so sure. Keep in mind that, in all countries today, the state already does this—without provoking discontent.

Take the UK, where people tend to spend five to six hours a day in leisure. (The average number is 6.1 hours a day for men, 5.5 for women.) Britons might imagine that they spend this time as they see fit. In fact, the state is lurking in the wings, quietly influencing what they do. According to the Office for National Statistics, the most popular leisure activity is consuming "mass media." This essentially means watching TV, with a little reading and music thrown in.[48] And it is true that people get to choose the TV channels and movies that they watch. But to own a TV in the UK in the first place, you have to pay an annual tax to fund the public broadcaster, the BBC. Their channels are the first you'll see when flicking through the options. The state gets to shape what those channels show, too: the BBC is required to "inform, educate, and entertain," lest it lose its ability to levy that tax.[49]

And how do Britons spend their spare time when they are not watching TV? There are a few hours a week playing sports or being outdoors, and about another hour a week doing cultural activities, like visiting museums or going to the theater. Once again, the state is here, quietly influencing things behind the scenes. In fact, in the UK, an entire government department—the Department for Digital, Culture, Media, and Sport—tries to shape how that time is spent. They make a range of interventions, like making sure all children have the chance to learn how to swim and ride a bike; providing entry to many of the best museums in the country completely for free; and banning the finest art in the country from being sold and taken overseas.[50] In fact, look to any area of our leisuring lives, and you are likely to find, if not a formal government department, then at least a network of publicly supported "trusts" and "foundations" and "bodies" that are gently cajoling us to adopt certain activities and abandon others.

There are also instances of what can be described as unintentional leisure policies. Pension systems are one example. Around the world, these are all based on the principle that leisure is something you should do in the twilight of life. But as Sarah O'Connor asks in the *Financial Times*, "if state support is going to allow everyone to have a period of leisure in their lives, why does it all have to be at the end?"[51] Already, in a world where life expectancy is improving, where workers would benefit from taking time off to retrain, and where people face substantial and irregular demands on their nonworking time (to bring up children, perhaps, or care for elderly relatives), it is striking that the state has decided to only really provide financial support for leisure once most of life is over.

Or consider the volunteering sector. In Britain today, about fifteen million people volunteer regularly—half as many people as there are in paid work.[52] (Andy Haldane, the chief economist at the Bank of England, estimates that the economic value of this volunteering in the UK is £50 billion a year, making it as valuable as the energy industry.)[53] But the sector does not operate in a vacuum: the state has a spread of programs and procedures designed to support it. These, too, can be seen as leisure policies of a sort, encouraging people to spend their spare time in a variety of particular activities for free.

As these examples show, there is already a variety of "leisure policies" in action today. However, right now they are a highly haphazard collection of minor intrusions, often accidental ones, on people's spare time. In a world with less work, this scattergun approach will no longer be appropriate. Societies will need to think far more deliberately, comprehensively, and coherently about their leisure policy.

This would be a radical change in direction. Leisure today is increasingly seen as a superfluity rather than a priority. The state, when looking to present itself as austere in its spending, often treats leisure activities as disposable, low-hanging fiscal fruit that can be cut and discarded with ease. In the United States, President Trump has tried to eliminate the National Endowment for the Arts, the Institute of Museum and Library Services, and the Corporation for Public Broadcasting.[54] In the UK, the number of public libraries has been cut by about 12 percent between 2010 and 2016.[55] (This decline has struck a national nerve: when the writer Philip Pullman spoke to a small gathering in Oxford about the fight against library closures, the online transcript of his speech became, in the words of one enthusiastic commentator, a "viral sensation."[56])

It is important, though, not to be overly prescriptive from today's armchairs about what particular pastimes a community might want to encourage its members to do. It is up to future generations to deliberate with one another about how to spend spare time in meaningful and purposeful ways. Attempts to forecast how people might spend their leisure time in the future have often turned out poorly. In 1939, for instance, the *New York Times* argued that TV would never catch on. "The problem with television is that people must sit and keep their eyes glued on a screen; the average American family hasn't time for it," the editorial confidently stated. "For this reason, if for no other, television will never be a serious competitor of broadcasting."[57] Needless to say, that prediction would turn out to be very wrong indeed.

A RETURN TO WORK

After a period of leisurely exploration, some people might conclude that for them, no activity can match work in providing a sense of fulfillment or direction. Even if their income comes from elsewhere, they may decide that a "job" is the only way to secure the meaning they seek.

One of my favorite poems, Alfred Tennyson's "Ulysses," is about this exact feeling. It tells the story of Odysseus, the Greek hero who, having won the Trojan war, takes a decade to return home from the battle. He is held up, among other obstacles, by lotus-eating people who try to sedate him, by the ghost of his dead mother who attempts to distract him, by a tribe of one-eyed giants who imprison him, and a different group of giants who try to eat him. It is, in short, a difficult commute home. And in the poem, Tennyson imagines how Odysseus would have felt upon returning from that adventure, having to settle down as an "idle king" on his throne again. His answer is that it would have been unbearably boring. Odysseus would struggle to "rust unburnish'd" but would want to "shine in use." And so, in Tennyson's poem, Odysseus prepares to set sail again, handing the throne over to his son, hoping that "some work of noble note, may yet be done / not unbecoming men that strove with Gods."[58] In a similar way, arriving at a world of technological unemployment, Odysseus-like people in the future might still want to attempt their own "work of noble note."

Until now, I have spoken about "a world with less work." What I have really meant, though, is a world with less *paid* work. Up to this point, there was little need to draw attention to that distinction. As we think about the future, though, it is important to look at this more carefully. Why? Because while we may be approaching a world with less paid work, there is no reason why it must be a world without any work at all. In the future, people who find themselves itching to keep at work despite having no economic reason to do so may hunt for roles that we would call "work" today—the only difference being that for them, this work would not come with a wage in the labor market that is large enough to live on.

What might these roles involve? Anything, in a sense, once the constraint of having to earn a wage to live on is removed. They might draw on tasks that machines could do better, but human beings want to do anyway in spite of that. That might sound inefficient, but if this work is done for noneconomic ends, in pursuit of purpose rather than productivity, then economic worries about "efficiency" are a mistake.

The existence of these Odysseus-like people creates another role for the state: to help those who want to work find such work. One possibility is that the state may actively create work for people to do. This is not as radical as it sounds: in fact, governments already do this on a huge scale. Seven of the top ten largest employers in the world are state-owned institutions—the US Department of Defense, the NHS, the State Grid Corporation of China, and the Indian Railways, to name a few. Even today, the idea of a "job guarantee" is already gathering favor and interest. In the United States, several Democratic candidates for the 2020 presidential election endorse the notion of offering everyone a job, and 52 percent of Americans support the policy. To put those numbers in context, the pollster responsible described it as "one of the most popular issues we've ever polled."[59]

The discussion about people who want to *work* in a world with less *paid work* does leave us with a conceptual puzzle. If people no longer directly rely upon that work for an income, is it still right to call it "work"—or is it leisure? In the Age of Labor, we have not had to worry too much about that distinction. Leisure is often defined simply as what people do when they are not at work, and work as what they do when they are not at leisure. In a world with less paid work, though, these definitions and boundaries become murkier. Is something only "work" if it is done for a wage? If so, that would suggest that housework, for instance, is not work. Is something only "work" if it is strenuous and hard, or perhaps slightly unpleasant? That would require us to say that people in paid but pleasurable work are at leisure (and that sports fans intensely watching television as their team loses are at work).

Philosophers have spent a lot of time trying to pin down this distinction.[60] But in practice, I am not sure how much it matters. When considering a future with less paid work, it is far more illuminating to think simply about *free time*. Some may want to spend some of that time doing things that look a lot like "leisure" today; others may incline toward more structured and directed roles, in the spirit of "work" in the past. My suspicion, though, is that what people choose to do will mostly *not* look like work today. Work is a source of meaning for some people at the moment not because work itself is special, but because our jobs are where we spend the majority of our lives. We can only find meaning in what we actually do—and freed up to spend our lives differently, we will find meaning elsewhere instead.

A ROLE FOR THE CONDITIONAL BASIC INCOME

And so we come back to the initial question of this chapter. What will people without work do with all of their free time? Part of the answer is that they might pursue more leisure. To that end, as we have seen, the state might want to step in and help them to use that time in a meaningful way. Another response is that some might want to retreat to some activity that looks more like work, though not in direct pursuit of a wage. And the state might want to support those ambitions, too.

But these two options are unlikely to be the full answer. In a world with less work, few societies will be able to allow those without a job to fill all their time with idleness, play, or unpaid work as they alone see fit. This is because, as noted earlier, any society that allows that is likely to fall apart. Today, social solidarity comes from a sense that everyone contributes to the collective pot through the paid work that they do and the taxes that they pay. Maintaining that solidarity in the future will require those without paid work to spend at least some of their time contributing to the pot in other, noneconomic ways.

This is what the "conditional basic income," the CBI I have proposed, is designed to support: it is a UBI, but one that requires its recipients to do something in return. If it is adopted, it means that, in the future, the daily lives of those without work are likely to be divided in two: not between leisure and paid work, but between activities that they choose and others that their community requires them to do.

We can speculate about what those required activities will be. Some communities, populated by people like Keynes and Russell, might be satisfied if those without work spend their time in pursuit of artistic and cultural ends: reading, writing, composing beautiful music, thinking deep thoughts. Others, channeling the ancient Greeks, might ask people to take their role as citizens more seriously; to engage in politics, to support local government, to ponder their obligations to others.[61] Beyond such recreational and political activities, I believe that educational, household, and caregiving activities will be recognized as important as well. No matter how capable machines become, we will, I imagine, want human beings to have a role in preparing others to live purposeful lives, and in supporting them during hard times and ill health.

This list, again, is speculative, and no doubt incomplete. In the end, it will be up to future communities to decide what counts as a contribution and what does not. Different societies will be drawn to different conclusions. But all of them, engaged in the same exercise, will be forced to say what they consider to be valuable and what not.

Today, that sense of value is overwhelmingly shaped by the market mechanism: a thing's value is the price that someone is willing to pay for it, and a worker's worth is the wage that they receive. For all its flaws, there is something extraordinary about the inexorable simplifying power of this mechanism. In the white heat of the market, the clash between people's infinite desires and the hard reality of satisfying them gets boiled down to a single number: a price.

Extraordinary, but still flawed. There are things we all recognize as being significant that have no price tag, and jobs we all think are important that receive little or no pay in return. Most caregiving, for instance, is unpaid.[62] In the United States, about 40 million family caregivers provide $500 billion of unpaid care to adults every year, two-thirds of it done by older women.[63] In the UK, around 6.5 million caregivers, again mostly women, provide unpaid care worth up to £100 billion.[64] Most housework, too, is unpaid. In the UK, the combined value of cooking, child care, laundry, and dealing with the general clutter of household chores is estimated to be about £800 billion—more than four times the size of the manufacturing sector.[65] Once again, it tends to be performed by women. A single number cannot capture all the different dimensions in which our sense of what matters can run.

In a world with less work, we will have an opportunity to repair this mismatch. President Obama hinted at this possibility in a set of parting reflections on the future of work. What we need to do, he said, is to begin to "reexamine what we value, what we are collectively willing to pay for—whether it's teachers, nurses, caregivers, moms or dads who stay at home, artists, all the things that are incredibly valuable to us right now but don't rank high on the pay totem pole."[66] And if we adopt a CBI, we will be driven to do exactly that: to take activities that the invisible hand of the labor market had marked down as worthless, and, with the visible hand of the community, to hold them up as being valuable and important. We will have a chance to allocate value through community recognition rather than through market wages. Fulfilling the requirements of a CBI may turn out to provide a sense of self-satisfaction not too different from that of bringing home a paycheck: the warm glow of earning one's keep, albeit earned in a different way.

THE MEANING-CREATING STATE

This final chapter is the most speculative in the book. But it contains two important lessons. The first is that if free time does become a bigger part of our lives, then it is likely to also become a bigger part of the state's role as well. Just as in the Age of Labor we have interventions designed to shape our working lives, in a world with less work we will need a set of tools to influence our free time, too. Those can include leisure policies, designed to help people spend their time in purposeful ways; opportunities for people who still want to "work" even if not for a wage; and requirements that people contribute to society, in return for the support that society provides. These are some possible directions. I am sure there will be more.

The second lesson is that work has meaning beyond the purely economic. That relationship does not always hold: for some people, their work is a source of income and nothing more. But for others, their work does provide a strong sense of purpose. They have economic identities, a sense of who they are, that is tightly rooted in the work that they do.

British coal miners show this very clearly. In the old mining city of Durham, in the north of England, the streets get closed to traffic once a year. The town fills up with crowds of miners and their supporters. There is music and marching, brass bands and celebratory songs; people carry huge banners above their heads decorated with the faces of historical mining heroes, and slogans like "unity" and "community" and "pride." This is a group of people whose identities are clearly anchored in their particular type of work. The Durham Miners' Gala, as the march is known, has been happening since 1871. But it is uncertain how much longer it will continue. At the end of 2015, the last deep coal mine in Britain, Kellingley Colliery, was closed, and the retired mining machines were buried in the old pits.[67] To me, the ceremony looked much like a set of religious funeral rites.

In a world with less work, the opportunity for economic identities like those of the British miners will shrivel. Instead, people will be forced to find noneconomic identities elsewhere. Today, there is a surge in identity politics: people's political tastes are increasingly shaped by their race or their faith or the place where they live. At times, I wonder whether this is in part a reaction to the insecurity of contemporary economic life, a retreat to a noneconomic source of meaning that feels sturdier and more reliable than the economic alternatives. But there are reasons to worry about these noneconomic identities, too. For one thing, people can be very bad at recognizing them. In Britain, the widespread failure among Remainers to imagine that Brexit could happen is a good example: there was a collective blindness to the fact that purpose in life might not only run in an economic dimension, that arguments about "trade" and "growth" might be answering questions that many people were simply not prioritizing.[68] What's more, the noneconomic identities that emerge may be quite unsavory. The recent rise of populist politics around the world, partly a response to economic insecurity, is an ominous example of this.

From these two lessons, a final role emerges for the Big State: as a meaning-creating state. As we approach a world with less work, a traditional source of purpose for many people will fall away and a gap will appear. New sources of purpose will emerge, not all of them benevolent. We may want a meaning-creating state to step in and, through interventions like leisure policies and the CBI, guide whatever floods in to fill work's place.

Of all the roles that I have set out for the Big State, this is the most unfamiliar. Today, we are used to our politicians acting as managers and technocrats whose role is to solve esoteric policy problems. We tend not to think of them as moral leaders. We do not expect them to guide us on what it means to live a flourishing life. But in a world with less work, we will need them to help us do this as well. "Towards what ultimate point is society tending by its industrial progress?" asked John Stuart Mill. "When the progress ceases, in what condition are we to expect that it will leave mankind?"[69] We may want a meaning-creating state to help us find answers.

Until now, modern political life has dodged philosophical questions like this. In the twentieth century, most societies agreed on the same goal: making the economic pie as large as it can be. And as Isaiah Berlin once wrote, "Where ends are agreed, the only questions left are those of means, and these are not political but technical, that is to say, capable of being settled by experts or machines like arguments between engineers or doctors."[70] Accordingly, fixated on this economic end, we have tended to turn to economists, the engineers of contemporary life, to tell us how to relentlessly grow that pie. In a world with less work, though, we will need to revisit the fundamental ends once again. The problem is not simply how to live, but how to live *well*. We will be forced to consider what it really means to live a meaningful life.

Epilogue

In 1941, Stefan Zweig sat at his desk in Brazil, writing. A decade before, he had been probably the most popular writer in Europe, with book sales that would make some of today's bestselling authors green with envy. By now, though, he had become an exile, yet another displaced Austrian Jew forced to flee his home. Sitting at that desk in Brazil, Zweig was writing his autobiography, *The World of Yesterday*. When he was growing up, he explains, everything—the buildings, the government, their way of life—seemed to everyone to be unshakable. He called it the "Golden Age of Security." As a child, he felt that this world would last forever. Alas, as we now know, it was not meant to be.[1]

Over the past decade, when reflecting on our future, I have often thought of Zweig sitting alone and working away at his book. To me, it seems that many of us have also grown up in an age of security—what I have called the Age of Labor. After the insanity and slaughter of the first half of the twentieth century, things took on a more predictable, calmer rhythm across much of the world, and the pursuit of paid work was an important part of that. The advice passed on to us from those further along in life was always the same. Our parents and teachers explained that if we kept our heads down, and tried hard at school or whatever else we chose to do, then a future of stable paid work would be waiting for us. As we got older we could expect to be paid more, and when we were old we could stop working and draw on the fruits of our labors. Life was all about work—preparing for it, doing it, retiring from it—and that seemed okay.

In this book, I have argued that our age of security, like Zweig's, is fated to come to an end. In the next hundred years, technological progress will make us more prosperous than ever before. Yet that progress will also carry us toward a world with less work for human beings. The economic problem that haunted our ancestors, that of making the economic pie large enough for everyone to live on, will fade away, and three new problems will emerge to take its place. First, the problem of inequality, of working out how to share this economic prosperity with everyone in society. Second, the problem of political power, of determining who gets to control the technologies responsible for this prosperity and on what terms. And third, the problem of meaning, of figuring out how to use this prosperity not just to live without work but to live well.

These problems are daunting and will be hard to solve. There will be fundamental disagreement about what we should do. Yet I am hopeful about the future. When we think about what lies ahead, it is important to look back as well, to remind ourselves of our three-hundred-thousand-year-old story so far and remember the challenges we have already overcome. Not that many generations ago, almost all human beings lived on or around the poverty line. The struggle for subsistence was the challenge that preoccupied most of mankind. Our generation has been fortunate to wake up in a world where people need not be condemned to that fate, where there is in principle enough economic prosperity for us all to keep ourselves and our families alive. The looming problems—of inequality, power, and meaning—are just the consequences of this unprecedented prosperity. They are the price we pay for the material abundance that some of us (though as yet not all of us) have been fortunate to enjoy. And in my view, it is a price worth paying.

In the twenty-first century, we will have to build a new age of security, one that no longer relies on paid work for its foundations. And we have to begin this task today. Although we cannot know exactly how long it will take to arrive at a world with less work for human beings to do, there are clear signs that we are on our way there. The problems of inequality, power, and meaning are not lurking in the distance, hidden out of sight in the remote future. They have already begun to unfold, to trouble and test our inherited institutions and traditional ways of life. It is up to us now to respond.

NOTES

PREFACE TO THE PAPERBACK EDITION

1. For the US, see Dominic Rushe, "US Economy Suffers Worst Quarter Since the Second World War as GDP Shrinks by 32.9%," *Guardian*, 30 July 2020; for the UK, see the ONS data in Delphine Strauss, "UK Economy Suffers Worst Slump in Europe in Second Quarter," *Financial Times*, 12 August 2020.
2. https://twitter.com/AndrewYang/status/1238095725721944065 (accessed 30 September 2020).
3. For the US, see "Coronavirus: US to Borrow Record $3tn as Spending Soars," BBC News, 4 May 2020; for the UK, see Chris Giles, "UK Public Finances Continue on Path to Record Peacetime Deficit," *Financial Times*, 25 September 2020.
4. Richard Henderson, "Big Tech Presents a Problem for Investors as Well as Congress," *Financial Times*, 1 August 2020.
5. "Apple More Valuable Than the Entire FTSE 100," BBC News, 2 September 2020.
6. For the UK, see Office for National Statistics, "Coronavirus and Depression in Adults, Great Britain: June 2020," 18 August 2020; for the US, see Paige Winfield Cunningham, "The Health 202: Texts to Federal Government Mental Health Hotline up Roughly 1,000 Percent," *Washington Post*, 4 May 2020.
7. Nir Jaimovich and Henry Siu, "Job Polarization and Jobless Recoveries," *Review of Economics and Statistics* 102, no. 1 (2020): 129–47. Whether the pattern is the same in other countries is less clear. See Joel Bilt, "Automation and Reallocation: Will COVID-19 Usher in the Future of Work?," *Canadian Public Policy* 46, no. S2 (2020), for evidence that it holds in Canada. Though also see Georg Graetz and Guy Michaels, "Is Modern Technology Responsible for Jobless Recoveries?," *American*

Economic Review 107, no. 5 (2017): 168–73, for evidence that it holds less strongly elsewhere.

8. Brigid Francis-Devine, Andrew Powell, and Niamh Foley, "Coronavirus: Impact on the Labor Market," House of Commons Library Brief Paper no. 8898 (12 August 2020). There were 28.02 million employees in the UK for April to June 2020.
9. EY, "Capital Strategies Being Rewritten as C-Suite Grapples with Immediate Impact of New Reality," 30 March 2020, published online at https://www.ey.com/en_gl/news/ (accessed 29 September 2020).
10. Ellen Daniel, "Brits More Positive About Technology Following Covid-19 Pandemic," Verdict, 18 June 2020, published online at https://www.verdict.co.uk/covid-19-technology-vodafone/ (accessed 29 September 2020); Vinous Ali, "Survey Results: Lockdown and Changing Attitudes Towards Tech," TechUK, 17 July 2020, published online at https://www.techuk.org/ (accessed 29 September 2020).
11. Paul Lynch and Daniel Wainwright, "Coronavirus: How GPs Have Stopped Seeing Most Patients in Person," BBC News, 11 April 2020.
12. Megan Brenan, "U.S. Workers Discovering Affinity for Remote Work," Gallup, 3 April 2020, says that "sixty-two percent of employed Americans currently say they have worked from home during the crisis"; CIPD, "Impact of COVID-19 on Working Lives: Findings from Our April 2020 Survey," 3 September 2020, published online at https://www.cipd.co.uk/ (accessed 29 September 2020), says that "of those still working, 61% were working remotely all the time."
13. Olga Khazan, "How the Coronavirus Could Create a New Working Class," *Atlantic*, 15 April 2020; Rakesh Kochhar and Jeffrey Passel, "Telework May Save U.S. Jobs in COVID-19 Downturn, Especially Among College Graduates," Pew Research Center, 6 May 2020.
14. For the UK, see Suzie Bailey and Michael West, "Ethnic Minority Deaths and COVID-19: What Are We to Do?," The Kings Fund, 30 April 2020; for the US, see CDC, "COVID-19 Hospitalization and Death by Race/Ethnicity," https://www.cdc.gov/coronavirus/2019-ncov/ (the data given here is from the 18 August 2020 update).
15. Tomaz Cajner, Leland Crane, Ryan Decker, John Grigsby, Adrian Hamins-Puertolas, Erik Hurst, Christopher Kurz, and Ahu Yildirmaz, "The U.S. Labor Market During the Beginning of the Pandemic Recession," Becker Friedman Institute Working Paper No. 2020-58 (2020).
16. See, for instance, McKinsey & Co., "Survey: US Consumer Sentiment During the Coronavirus Crisis," 28 August 2020.
17. See David Autor and Elisabeth Reynolds, "The Nature of Work After the COVID Crisis: Too Few Low-Wage Jobs," The Hamilton Project, Essay 2020-14, July 2020. The authors make a similar argument to the one that follows.

INTRODUCTION

1. This story, as we shall see, travels under various names and in various forms. To see it referred to as the "Great Manure Crisis," consider, for instance, Brian Groom, "The Wisdom of Horse Manure," *Financial Times*, 2 September 2013; and Stephen Davies, "The Great Horse-Manure Crisis of 1894," September 2004, https://admin.fee.org/files/docLib/547_32.pdf (accessed January 2019).

2. Maxwell Lay, *Ways of the World: A History of the World's Roads and of the Vehicles That Used Them* (New Brunswick, NJ: Rutgers University Press, 1992), p. 131.
3. Vic Sanborn, "Victorian and Edwardian Horse Cabs by Trevor May, a Book Review," 17 November 2009, https://janeaustensworld.wordpress.com/tag/horse-drawn-cabs/ (accessed February 2019); Elizabeth Kolbert, "Hosed: Is There a Quick Fix for the Climate?," *New Yorker*, 16 November 2009; Davies, "Great Horse-Manure Crisis."
4. Jennifer Lee, "When Horses Posed a Public Health Hazard," *New York Times*, 9 June 2008.
5. Ted Steinberg, *Down to Earth: Nature's Role in American History* (New York: Oxford University Press, 2002), p. 162.
6. Kolbert, "Hosed"; Davies, "Great Horse-Manure Crisis"; Eric Morris, "From Horse Power to Horsepower," *ACCESS Magazine* 30 (Spring 2007).
7. Lee, "When Horses Posed."
8. Steven Levitt and Stephen Dubner, *Superfreakonomics* (New York: HarperCollins, 2009).
9. "The Horse Plague," *New York Times*, 25 October 1872; Sean Kheraj, "The Great Epizootic of 1872–73: Networks of Animal Disease in North American Urban Environments," *Environmental History* 23:3 (2018).
10. Steinberg, *Down to Earth*, p. 162.
11. "The Future of Oil," *Economist*, 26 November 2016.
12. But, as the double entendre suggests, not without embellishment. See, for instance, Rose Wild, "We Were Buried in Fake News as Long Ago as 1894," *Sunday Times*, 13 January 2018.
13. See, for instance: Wassily Leontief, "Technological Advance, Economic Growth, and the Distribution of Income," *Population and Development Review* 9, no. 3 (1983): 403–10; "Is Technological Unemployment Inevitable?," *Challenge* 22, no. 4 (1979): 48–50; Wassily Leontief, "National Perspective: The Definition of Problems and Opportunities" in *The Long-Term Impact of Technology on Employment and Unemployment: A National Academy of Engineering Symposium, 30 June 1983* (Washington, DC: National Academy Press, 1983).
14. Georg Graetz and Guy Michaels, "Robots at Work," *Review of Economics and Statistics* 100, no. 5 (2018): 753–68; Aaron Smith and Monica Anderson, "Automation in Everyday Life," Pew Research Center, 4 October 2017, http://www.pewinternet.org/2017/10/04/automation-in-everyday-life/ (accessed August 2018).
15. Grace, Katja, John Salvatier, Allan Dafoe, et al., "When Will AI Exceed Human Performance? Evidence from AI Experts," *Journal of Artificial Intelligence Research* 62 (2018): 729–54.
16. Nicholas Bloom, Chad Jones, John Van Reenan, and Michael Webb, "Ideas Aren't Running Out, but They Are Getting More Expensive to Find," Voxeu.org, 20 September 2017.
17. Daniel Susskind, "Technology and Employment: Tasks, Capabilities and Tastes," DPhil. diss. (Oxford University, 2016); Daniel Susskind and Richard Susskind, *The Future of the Professions* (Oxford: Oxford University Press, 2015).
18. "Materially prosperous" means taking into account things like their food and clothing, housing and heating. From Gregory Clark, *A Farewell to Alms* (Princeton, NJ: Princeton University Press, 2007), p. 1.

19. The "$80.7 trillion pie" is the $80.738 trillion global GDP in current US dollars for 2017, from the World Bank, https://data.worldbank.org/indicator/NY.GDP.MKTP.CD. The "7.53 billion people" is the global population in 2017, also from the World Bank, https://data.worldbank.org/indicator/SP.POP.TOTL?page=2. Joseph Stiglitz does this calculation as well, when thinking about John Maynard Keynes and his prophecies. See Joseph Stiglitz, "Toward a General Theory of Consumerism: Reflections on Keynes's Economic Possibilities for Our Grandchildren" in Lorenzo Pecchi and Gustavo Piga, eds., *Revisiting Keynes: Economics Possibilities for Our Grandchildren* (Cambridge, MA: MIT Press, 2008).
20. John Kenneth Galbraith, *The Affluent Society* (London: Penguin Books, 1999), p. 4.
21. Charlotte Curtis, "Machines vs. Workers," *New York Times*, 8 February 1983.
22. See the Introduction to Karl Popper, *The Open Society and Its Enemies*, vol. 1: *The Age of Plato* (London: Routledge, 1945).

1. A HISTORY OF MISPLACED ANXIETY

1. See James Lovelock, *Novacene* (London: Allen Lane, 2019) p. 1; and Yuval Noah Harari, *Sapiens*, (London: Harvill Secker, 2011), chap. 1.
2. The world population grew in that time, too, which is why the second number is so much larger. Data from Angus Maddison, *The World Economy: A Millennial Perspective*, http://www.theworldeconomy.org/ (2006).
3. In a recent disagreement, one economist accused two others of presenting "a stream of assertions and anecdotes"; the pair responded that their accuser was throwing "a lot of mud, hoping that some if it would stick." See Jeffrey Sachs, "Government, Geography, and Growth," *Foreign Affairs*, September/October 2012; and the response from Daron Acemoglu and James Robinson, "Response to Jeffrey Sachs," 21 November 2012, http://whynationsfail.com/blog/2012/11/21/response-to-jeffrey-sachs.html.
4. Data from Angus Maddison, *Historical Statistics of the World Economy*, http://www.ggdc.net/maddison/oriindex.htm (2010).
5. Britain's rate of growth, though, is again contested among economists. For instance, see Pol Antràs and Hans-Joachim Voth, "Factor Prices and Productivity Growth During the British Industrial Revolution," *Explorations in Economic History* 40 (2003): 52–77.
6. Britain's rate of productivity growth during the Industrial Revolution is also the subject of debate among economists. For instance, see ibid.
7. Joel Mokyr, "Technological Inertia in Economic History," *Journal of Economic History* 52, no. 2 (1992): 325–38, n. 17; David Weil, *Economic Growth*, 3rd ed. (London: Routledge, 2016), p. 292.
8. Eric Hobsbawm, *Industry and Empire* (London: Penguin, 1999), p. 112.
9. This story is from Robert Allen, "The Industrial Revolution in Miniature: The Spinning Jenny in Britain, France, and India," Oxford University Working Paper No. 375 (2007).
10. *Athenaeum* 18 July 1863 (no. 1864), p. 75.
11. See John Kay, "What the Other John Kay Taught Uber About Innovation," *Financial Times*, 26 January 2016. In John Kay, "Weaving the Fine Fabric of Success," https://www.johnkay.com/, 2 January 2003, Kay questions quite how reliable this tale really

is, writing that "Kay's own account tells us he fled to France to escape unemployed weavers: more likely he fled to France to escape his creditors."

12. See http://statutes.org.uk/site/the-statutes/nineteenth-century/1812-52-geo-3-c-16-the-frame-breaking-act/; and http://statutes.org.uk/site/the-statutes/nineteenth-century/1813-54-geo-3-cap-42-the-frame-breaking-act/.
13. Daron Acemoglu and and James Robinson, *Why Nations Fail* (London: Profile Books, 2012), pp. 182–83. Perhaps we should be suspicious of her true motivations; this was, in fact, the queen's second refusal; the first was on the less lofty grounds that Lee's machines created stockings that were coarser than the silk alternatives she liked from Spain. See Marjorie Garber, *Vested Interests: Cross-Dressing and Cultural Anxiety* (New York: Routledge, 2012), p. 393, n. 6.
14. See Anni Albers, *On Weaving* (Princeton, NJ: Princeton University Press, 2017), p. 15; and Eric Broudy, *The Book of Looms* (Hanover, NH: University Press of New England, 1979), p. 146, for two competing accounts of the murder. Others say Möller was drowned in the Vistula River by a rabble of weavers who feared the competition. I first came across this story in Ben Seligman, *Most Notorious Victory: Man in an Age of Automation* (New York: Free Press, 1966).
15. The first edition was published in 1817, the third edition, with the new chapter, in 1821. See David Ricardo, *Principles of Political Economy and Taxation* (New York: Prometheus Books, 1996).
16. "Automation and Anxiety," *Economist*, 25 June 2016; and Louis Stark, "Does Machine Displace Men in the Long Run?," *New York Times*, 25 February 1940.
17. For President Obama's farewell speech, see Claire Cain Miller, "A Darker Theme in Obama's Farewell: Automation Can Divide Us," *New York Times*, 12 January 2017. President Kennedy gave his speech at the AFL-CIO Convention, Grand Rapids, Michigan, 7 June 1960; see https://www.jfklibrary.org/.
18. Stephen Hawking, "This Is the Most Dangerous Time for Our Planet," *Guardian*, 1 December 2016.
19. See "World Ills Laid to Machine by Einstein in Berlin Speech," *New York Times*, 22 October 1931. In David Reichinstein, *Albert Einstein: A Picture of His Life and His Conception of the World* (Prague: Stella Publishing House, 1934), p. 96, the account of that speech reveals that Einstein was worrying, in part, about technological unemployment.
20. For instance: "March of the Machine Makes Idle Hands," 26 February 1928; "Technological Unemployment," 12 August 1930; "Does Man Displace Men in the Long Run?," 25 February 1940; "'Revolution' Is Seen in 'Thinking Machines,'" 17 November 1950; "Newer and Newer and Newer Technology, with Little Unemployment," 6 March 1979; "A Robot Is After Your Job," 3 September 1980; "If Productivity's Rising, Why Are Jobs Paying Less?," 19 September 1993; "A 'Miracle,' But Maybe Not Always a Blessing," 25 February 2001.
21. Robert Tombs, *The English and Their History* (London: Penguin Books, 2015), pp. 377–78.
22. Ibid., p. 378.
23. From tables A49 and A50, version 3.1 of Ryland Thomas and Nicholas Dimsdale, "A Millennium of UK Data," Bank of England OBRA data set (2017), http://www.bankofengland.co.uk/research/Pages/onebank/threecenturies.aspx. I use the adapted Feldstein data in the former for 1760–1855, and the latter for 1855 to 1900. After 1855, the data is for the UK, rather than Britain.

24. Tyler Cowen, "Industrial Revolution Comparisons Aren't Comforting," *Bloomberg View*, 16 February 2017.
25. John Maynard Keynes, *Essays in Persuasion* (New York: W. W. Norton, 1963), pp. 368–69.
26. Data extracted from OECD. Stat, https://stats.oecd.org/, on April 2019.
27. See OECD (2017), https://data.oecd.org/emp/hours-worked.htm (accessed 1 May 2018).
28. From Max Roser, "Working Hours," https://ourworldindata.org/working-hours (accessed July 2018). Int-$ (the "international dollar") is a hypothetical currency that tries to take account of different price levels across different countries.
29. For instance, Daron Acemoglu and Pascual Restrepo, "Artificial Intelligence, Automation and Work" in Ajay Agrawal, Joshua Gans, and Avi Goldfarb, eds., *Economics of Artificial Intelligence* (Chicago: Chicago University Press, 2018).
30. Dayong Wang, Aditya Khosla, Rishab Gargeya, et al., "Deep Learning for Identifying Metastatic Breast Cancer," https://arxiv.org, arXiv:1606.05718 (2016).
31. Maura Grossman and Gordon Cormack, "Technology-Assisted Review in e-Discovery Can Be More Effective and More Efficient than Exhaustive Manual Review," *Richmond Journal of Law and Technology* 17, no. 3 (2011).
32. Data is from Maddison, *Historical Statistics*. It may seem odd to speak of the "US economy" in 1700—it did not exist then. This is the case for other countries in the data set, too. In doing so, I am relying on Maddison's own classifications. In the US case, this classification encompasses the British Empire's colonial territories.
33. Lawrence Summers, "The 2013 Martin Feldstein Lecture: Economic Possibilities for Our Children," *NBER Reporter* 4 (2013).
34. David Autor, "The Limits of the Digital Revolution: Why Our Washing Machines Won't Go to the Moon," Social Europe (2015), https://www.socialeurope.eu/.
35. Quoted in Blaine Harden, "Recession Technology Threaten Workers," *Washington Post*, 26 December 1982.
36. Stephen Broadberry, Bruce Campbell, Alexander Klein, *et al.*, *British Economic Growth, 1270–1870* (Cambridge: Cambridge University Press, 2015), p. 194, table 5.01.
37. The 1900 statistic is from David Autor, "Why Are There Still So Many Jobs? The History and Future of Workplace Automation," *Journal of Economics Perspectives* 29, no. 3 (2015): 3–30; the "today" statistic is 1.5 percent in 2016, from the US Bureau of Labor Statistics, https://www.bls.gov/emp/ep_table_201.htm.
38. The "quarter" is 26.4 percent, from the Federal Reserve Bank of St. Louis—https://fred.stlouisfed.org/series/USAPEFANA—and the "tenth" is 9 percent from the National Association of Manufacturers, "Top 20 Facts About Manufacturing," http://www.nam.org/Newsroom/Top-20-Facts-About-Manufacturing/.
39. US Bureau of Labor Statistics, https://www.bls.gov/emp/tables/employment-by-major-industry-sector.htm (accessed August 2019).
40. See, for instance, Daron Acemoglu, "Advanced Economic Growth: Lecture 19: Structural Change," delivered at MIT, 12 November 2017.
41. Jesus Felipe, Connie Bayudan-Dacuycuy, and Matteo Lanzafame, "The Declining Share of Agricultural Employment in China: How Fast?" *Structural Change and Economic Dynamics* 37 (2016): 127–37.
42. Estimate of "males in agriculture" in 1900–1909 is 810,000 in Gregory Clark, "The Agricultural Revolution and the Industrial Revolution: England, 1500–1912," unpublished

ms. (University of California, Davis, 2002). The National Health Service of England and Wales employed about 1.2 million people in 2017; see https://digital.nhs.uk/.

43. David Autor, "Polanyi's Paradox and the Shape of Employment Growth" in "Re-evaluating Labor Market Dynamics: A Symposium Sponsored by the Federal Reserve Bank of Kansas City. Jackson Hole, Wyoming, August 21–23, 2014" (2015) makes a similar point: "It is unlikely . . ." (p. 162).
44. Autor, "Why Are There Still So Many Jobs?," p. 5.
45. Bernardo Bátiz-Lazo, "A Brief History of the ATM," *Atlantic*, 26 March 2015.
46. Ibid.
47. These numbers and those that follow are from James Bessen, "Toil and Technology," *IMF Financial and Development* 51, no. 1 (2015). For the "20 percent," see chart 1—from approximately 500,000 tellers in the late 1980s, to approximately 600,000 in the late 2000s.
48. Lots of other economists have explored this problem. See Autor, "Why Are There Still So Many Jobs?"; and Bessen, "Toil and Technology," for instance. James Surowiecki, "Robots Won't Take All Our Jobs," *Wired*, 12 September 2017, is another.

2. THE AGE OF LABOR

1. See, for instance, Daron Acemoglu, "Technical Change, Inequality, and the Labor Market," *Journal of Economic Literature* 40, no. 1 (2002): 7–72.
2. David Autor, Lawrence Katz, and Alan Krueger, "Computing Inequality: Have Computers Changed the Labor Market?" *Quarterly Journal of Economics* 133, no. 1 (1998): 1169–1213.
3. The totals were 56.6 in 2000, 61.9 in 2001. World Bank data retrieved from "Personal Computers (per 100 People)" from https://datamarket.com/ (accessed July 2018).
4. William Nordhaus, "Two Centuries of Productivity Growth in Computing," *Journal of Economic History* 67, no. 1 (2007): 128–59.
5. From ibid., Data Appendix. Thank you to William Nordhaus for sharing his revised data with me.
6. Daron Acemoglu and David Autor, "Skills, Tasks and Technologies: Implications for Employment and Earnings" in David Card and Orley Ashenfelter, eds., *Handbook of Labor Economics*, vol. 4, pt. B (North-Holland: Elsevier, 2011), pp. 1043–171. The percentage difference between two variables is approximately equal to 100 times the exponential of the difference between the logs of those two variables, minus 1. Here, $100(e^{0.68}-1) = 97.4$, to three significant figures.
7. See, for instance, Eli Berman, John Bound, and Stephen Machin, "Implications of Skill-Biased Technological Change: International Evidence," *Quarterly Journal of Economics* 113, no. 4 (1998): 1245–79.
8. Acemoglu and Autor, "Skills, Tasks and Technologies," data from figure 1.
9. Data is figure 6 in David Autor, "Skills, Education, and the Rise of Earnings Inequality Among the 'Other 99 Percent,'" *Science* 344, no. 6186 (2014): 843–51.
10. From Max Roser and Mohamed Nagdy, "Returns to Education," https://ourworldindata.org/returns-to-education (accessed 1 May 2018).
11. See Daron Acemoglu, "Technical Change, Inequality, and the Labor Market," *Journal of Economic Literature* 40, no. 1 (2002): 7–72.

12. For England, see Alexandra Pleijt and Jacob Weisdorf, "Human Capital Formation from Occupations: The 'Deskilling Hypothesis' Revisited," *Cliometrica* 11, no. 1 (2017): 1–30. A similar story unfolded in the United States; see Kevin O'Rourke, Ahmed Rahman, and Alan Taylor, "Luddites, the Industrial Revolution, and the Demographic Transition," *Journal of Economic Growth* 18, no. 4 (2013): 373–409.
13. Quoted in Ben Seligman, *Most Notorious Victory: Man in an Age of Automation* (New York: Free Press, 1966), p. 11.
14. Joel Mokyr, *The Lever of Riches: Technological Creativity and Economic Progress* (New York: Oxford University Press, 1990), p. 137, quoted in O'Rourke et al., "Luddites."
15. Economists captured this story in a mathematical expression called the "constant elasticity of substitution production function." In economics, a "production function" tells you how different types of input (workers and machines, for instance) come together to produce some output. And this particular version was characterized by a "constant elasticity," meaning that a percentage change in the relative price of two inputs would always cause a constant percentage change in the use of those inputs. In this model, new technologies could only ever complement workers. See Acemoglu and Autor, "Skills, Tasks and Technologies," p. 1096, for an account of the "canonical model," and p. 1105, implication 2, for a statement of the fact that any kind of technological progress in the canonical model leads to an increase in absolute wages of both types of labor.
16. This is an edited version of figure 3.1 in OECD, *OECD Employment Outlook 2017* (Paris: OECD Publishing, 2017).
17. See Autor, "Polanyi's Paradox and the Shape of Employment Growth."
18. The exact nature of polarization depends upon the country. See, for instance, Maarten Goos, Alan Manning, and Anna Salomons, "Explaining Job Polarization: Routine-Biased Technological Change and Offshoring," *American Economic Review* 104, no. 8 (2014): 2509–26; David Autor, "The Polarization of Job Opportunities in the U.S. Labor Market: Implications for Employment and Earnings," Center for American Progress (April 2010); David Autor and David Dorn, "The Growth of Low-Skill Service Jobs and the Polarization of the US Labor Market," *American Economic Review* 103, no. 5 (2013): 1553–97; and Maarten Goos and Alan Manning, "Lousy and Lovely Jobs: The Rising Polarization of Work in Britain," *Review of Economics and Statistics* 89, no. 1 (2007): 119–33.
19. For the 0.01 percent statistic, see Emmanuel Saez, "Striking It Richer: The Evolution of Top Incomes in the United States," published online at https://eml.berkeley.edu/~saez/ (2016). For the "super-star bias," see Erik Brynjolfsson, "AI and the Economy," lecture at the Future of Life Institute, 1 July 2017.
20. See Acemoglu and Autor, "Skills, Tasks and Technologies," p. 1070, n. 25.
21. The classic statement of the ALM hypothesis is David Autor, Frank Levy, and Richard Murnane, "The Skill Content of Recent Technological Change: An Empirical Exploration," *Quarterly Journal of Economics* 118, no. 4 (2003): 129–333. This early paper is focused on explaining skills-biased technological change. In the years to come, that would change and the focus would turn to using the ALM hypothesis to explain polarization instead.
22. I explore this intellectual history in Susskind, "Technology and Employment," chap. 1.
23. This distinction is from Michael Polanyi, *The Tacit Dimension* (Chicago: Chicago University Press, 1966). To see this distinction in action, think of a great doctor.

Imagine asking how she makes such perceptive medical diagnoses. She might be able to give you a few hints, but ultimately she would struggle to explain herself. As Polanyi himself put it, very often "we can know more than we can tell." Economists called this constraint on automation "Polanyi's Paradox."

24. This is the language I used in my TED talk entitled "Three Myths About the Future of Work (and Why They Are Wrong)," March 2018. See David Autor, Frank Levy, and Richard Murnane, "The Skill Content of Recent Technological Change: An Empirical Exploration," *Quarterly Journal of Economics* 118, no. 4 (2003): 129–333.
25. Autor, "Polanyi's Paradox and the Shape of Employment Growth." These economists had company thinking this way, too. Wassily Leontief wrote, back in 1983, how "any worker who now performs his task by following specific instructions can, in principle, be replaced by a machine." See Leontief, "National Perspective," p. 3. Leontief, though, was quite a bit more pessimistic about the future than Autor.
26. Goos and Manning in "Lousy and Lovely Jobs" were perhaps the first to put the ALM hypothesis to use in this way.
27. Hans Moravec, *Mind Children* (Cambridge, MA: Harvard University Press, 1988).
28. The origins of this quotation are contested. The earliest recorded version of it comes from the Nobel Prize–winning economist Paul Samuelson, but Samuelson himself later attributed it to Keynes. See http://quoteinvestigator.com/2011/07/22/keynes-change-mind/.
29. Carl Frey and Michael Osborne, "The Future of Employment: How Susceptible Are Jobs to Computerisation?," *Technological Forecasting and Social Change* 114 (January 2017): 254–80.
30. McKinsey Global Institute, "A Future That Works: Automation, Employment, and Productivity," January 2017.
31. James Bessen, an economist at Boston University, was perhaps the first to note this.
32. Autor, "Why Are There Still So Many Jobs?"
33. Ibid.
34. There are lots of examples of this. Consider, for instance: IMF, *World Economic Outlook* (2017); World Bank, "World Development Report 2016: Digital Dividends," 14 January 2016; Irmgard Nübler, "New Technologies: A Jobless Future or Golden Age of Job Creation?," International Labour Office, Working Paper No. 13 (November 2016); Executive Office of the President, "Artificial Intelligence, Automation, and the Economy," December 2016.
35. Fergal O'Brien and Maciej Onoszko, "Tech Upheaval Means a 'Massacre of the Dilberts' BOE's Carney Says," *Bloomberg*, 13 April 2018.
36. Scott Dadich, "Barack Obama, Neural Nets, Self-Driving Cars, and the Future of the World," *Wired*, November 2016.
37. UBS, "Intelligence Automation: A UBS Group Innovation White Paper" (2017); PwC, "Workforce of the Future: The Competing Forces Shaping 2030" (2018); Deloitte, "From Brawn to Brains: The Impact of Technology on Jobs in the UK" (2015).
38. "Automation and Anxiety," *Economist*; and Elizabeth Kolbert, "Our Automated Future," *New Yorker*, 19 and 26 December 2016.
39. Isaiah Berlin, *Two Concepts of Liberty* (Oxford: Clarendon Press, 1958), p. 4. Quoting Heinrich Heine, a German poet, he writes how "philosophical concepts nurtured in the stillness of a professor's study could destroy a civilization."

3. THE PRAGMATIST REVOLUTION

1. Homer, *Iliad*, book 18, lines 370–80, http://www.perseus.tufts.edu/.
2. That Daedalus's sculptures were so lifelike is alluded to in Plato's work *Euthyphro*; see the http://www.perseus.tufts.edu version, p. 11, and accompanying notes.
3. It is said that Archytas, Plato's close friend, crafted the world's first robot—a steam-powered pigeon that could fly unaided through the air.
4. These examples, and many others, are set out in Nils J. Nilsson, *The Quest for Artificial Intelligence* (New York: Cambridge University Press, 2010).
5. For the cart and robot, see http://www.da-vinci-inventions.com/ (accessed 8 May 2018); for the lion, see Stelle Shirbon, "Da Vinci's Lion Prowls Again After 500 Years," Reuters, 14 August 2009.
6. Gaby Wood, *Living Dolls* (London: Faber and Faber, 2002), p. 35.
7. See Tom Standage, *The Turk* (New York: Berkley Publishing Group, 2002); and Wood, *Living Dolls*, pp. 79 and 81. The secret was only revealed in 1834, sixty-five years after the Turk was built—Jacques-Francois Mouret, one of the "directors" who had hidden in the machine, sold his secret to a newspaper.
8. Wood, *Living Dolls*, p. 35
9. Alan Turing, "Lecture to the London Mathematical Society," 20 February 1947; archived at https://www.vordenker.de/downloads/turing-vorlesung.pdf (accessed July 2018).
10. Alan Turing, "Intelligent Machinery: A Report by A. M. Turing," National Physical Laboratory (1948); archived at https://www.npl.co.uk (accessed July 2018).
11. See Grace Solomonoff, "Ray Solomonoff and the Dartmouth Summer Research Project in Artificial Intelligence" (no date), http://raysolomonoff.com/dartmouth/dartray.pdf.
12. John McCarthy, Marvin Minsky, Nathaniel Rochester, and Claude Shannon, "A Proposal for the Dartmouth Summer Research Project on Artificial Intelligence," 31 August 1955.
13. Susskind and Susskind, *Future of the Professions*, p. 182.
14. Marvin Minsky, "Neural Nets and the Brain Model Problem," PhD diss. (Princeton University, 1954).
15. Alan Newell and Herbert Simon, "GPS, A Program That Simulates Human Thought" in H. Billing, ed., *Lernende automaten* (Munchen: R. Oldenbourgh, 1961).
16. Here, I have in mind the widespread attempts to detect edges in images and, from those, represent objects as simple line drawings. David Marr, a British neuroscientist and psychologist, called these line drawings "primal sketches." This, he argued, was how human beings interpret the world. See Nilsson, *Quest for Artificial Intelligence*, chap. 9; and David Marr, *Vision: A Computational Investigation into the Human Representation and Processing of Visual Information* (London: MIT Press, 2010).
17. Alan Turing, "Intelligent Machinery, A Heretical Theory," *Philosophia Mathematica* 3, no. 4 (1996): 156–260.
18. Nilsson, *Quest for Artificial Intelligence*, p. 62.
19. John Haugeland, *Artificial Intelligence: The Very Idea* (London: MIT Press, 1989), p. 2.
20. Not all were happy with this route, though. The philosopher Hubert Dreyfus, a perennial AI pessimist, railed against colleagues who thought "they could program computers to

be intelligent like people." Quoted in William Grimes, "Hubert L. Dreyfus, Philosopher of the Limits of Computers, Dies at 87," *New York Times*, 2 May 2017.

21. Douglas Hofstadter, *Gödel, Escher, Bach: An Eternal Golden Braid* (London: Penguin, 2000), p. 579.
22. Daniel Crevier, *AI: The Tumultuous History of the Search for Artificial Intelligence* (New York: Basic Books, 1993), pp. 48 and 52.
23. "330 million moves" from Murray Campbell, A. Joseph Hoane Jr., and Feng-hsiung Hsu, "Deep Blue," *Artificial Intelligence* 134 (2002), 57–82; "a hundred possible moves" from the introduction to the revised edition of Hubert Dreyfus, *What Computers Can't Do: The Limits of Artificial Intelligence* (New York: Harper & Row, 1979), p. 30.
24. Data is from an ImageNet presentation on the 2017 challenge; see http://image-net.org/challenges/talks_2017/ILSVRC2017_overview.pdf (accessed July 2018). The Electronic Frontier Foundation lists the winning systems in a similar chart, and also plots the human error rate; see https://www.eff.org/ai/metrics#Vision (accessed July 2018). For an overview of the challenge, see Olga Russakovsky, Jia Deng, Hao Su, et al., "ImageNet Large Scale Visual Recognition Challenge," *International Journal of Computer Vision* 115, no. 3 (2015): 211–52.
25. Quoted in Susskind and Susskind, *Future of the Professions*, p. 161.
26. Not all researchers changed direction of travel in this way, though. Marvin Minsky in fact made the opposite move, moving from bottom-up approaches to AI, to top-down ones instead; see https://www.youtube.com/watch?v=nXrTXiJM4Fg.
27. Warren McCulloch and Walter Pitts, for instance, the first to construct one in 1943, were trying to describe "neural events" in the brain as "proposition logic" on paper. See Warren McCulloch and Walter Pitts, "A Logical Calculus of the Ideas Immanent in Nervous Activity," *Bulletin of Mathematical Biophysics* 5 (1943): 115–33.
28. This is why it does not matter that, as yet, we haven't even managed to simulate the workings of a worm's brain, with only about 302 neurons, never mind a human one, with around 100,000,000,000. Hannah Fry, *Hello World: How to Be Human in the Age of the Machine* (London: Penguin, 2018), p. 13.
29. On the fourth move, that rises to 280 billion times. For the number of possible chess positions after *n* "plies" (i.e., a move by only one player), see http://oeis.org/A019319 and http://mathworld.wolfram.com/Chess.html. For go moves, there are approximately: $361 \times 360 = 129{,}960$ possibilities after one move each; $361 \times 360 \times 359 \times 358$ after two moves each; and $361 \times 360 \times 359 \times 358 \times 357 \times 356$ after three moves each. (These are only rough calculations, since they assume that, on each move, a stone can be placed on any unoccupied point on the board—though there are conceivable situations in which one of those moves might be illegal.)
30. IBM described how "Deep Blue's 'chess knowledge' has improved since last year. Working with international grandmaster Joel Benjamin, the development team has spent the past several months educating Deep Blue about some of the finer points of the game." See https://www.research.ibm.com/deepblue/meet/html/d.3.3a.html (accessed 1 August 2017).
31. David Silver, Aja Huang, Chris Maddison, et al., "Mastering the Game of Go with Deep Neural Networks and Tree Search," *Nature* 529 (2016): 484–89; and David Silver, Julian Schrittwieser, Karen Simonyan, et al., "Mastering the Game of Go Without Human Knowledge," *Nature* 550 (2017): 354–59.

32. Matej Marovčík, Martin Schmid, Neil Burch, et al., "Deep Stack: Expert-Level Artificial Intelligence in Heads-Up No-Limit Poker," *Science* 356, no. 6337 (2017): 508–13.
33. Noam Brown and Tuomas Sandholm, "Superhuman AI for Multiplayer Poker," *Science* (2019), https://science.sciencemag.org/content/early/2019/07/10/science.aay2400.
34. Newell and Simon, "GPS."
35. Dreyfus, *What Computers Can't Do*, p. 3.
36. Consider, for instance, Marvin Minsky's definition, that AI "is the science of making machines do things that would require intelligence if done by men." In Minsky, *Semantic Information Processing* (Cambridge, MA: MIT Press, 1968), p. 5.
37. From Hilary Putnam, "Much Ado About Not Very Much," *Daedalus* 117, no. 1 (1988): 269–81.
38. See Haugeland, *Artificial Intelligence*, p. 5; and Margaret Boden, *Philosophy of Artificial Intelligence* (Oxford: Oxford University Press, 1990), p. 1.
39. Ibid.
40. Cade Metz, "A.I. Researchers Are Making More Than $1 Million, Even at a Nonprofit," *New York Times*, 19 April 2018.
41. Yaniv Leviathan and Yossi Matias, "Google Duplex: An AI System for Accomplishing Real-World Tasks over the Phone," 8 May 2018, https://ai.googleblog.com/ (accessed August 2018).
42. See William Paley, *Natural Theology* (Oxford: Oxford University Press, 2008); and Genesis 1:27, http://biblehub.com/genesis/1-27.htm.
43. This line, and a similar sentiment of argument to the one I am making, can be found in both Daniel Dennett, *From Bacteria to Bach and Back* (London: Allen Lane, 2017) and Dennett, "A Perfect and Beautiful Machine: What Darwin's Theory of Evolution Reveals About Artificial Intelligence," *Atlantic*, 22 June 2012. In collecting my own thoughts over the years, I have been influenced by the way that Dennett thinks about the relationship between evolution by natural selection and machine learning.
44. Charles Darwin, *On the Origin of Species* (London: Penguin Books, 2009), p. 401.
45. Richard Dawkins, *The Blind Watchmaker* (London: Penguin Books, 2016), p. 9.
46. Again, Daniel Dennett makes similar arguments in his work. See Dennett, *From Bacteria to Bach* and "A Perfect and Beautiful Machine."

4. UNDERESTIMATING MACHINES

1. Joseph Weizenbaum, "ELIZA—A Computer Program for the Study of Natural Language Communication Between Man and Machine," *Communications of the ACM* 9, no. 1 (1966): 36–45. ELIZA was named after Eliza Doolittle, a flower girl with a thick Cockney accent in George Bernard Shaw's play *Pygmalion*, who is "taught to 'speak' increasingly well" and swiftly enters into high-society London as a result.
2. See Joseph Weizenbaum, *Computer Power and Human Reason* (San Francisco: W. H. Freeman, 1976) for a full account of ELIZA and its consequences.
3. Ibid., p. 6.
4. Quoted in Bruce Weber, "Mean Chess-Playing Computer Tears at Meaning of Thought," *New York Times*, 19 February 1996.
5. Douglas Hofstadter, "Just Who Will Be We, in 2493?," Indiana University, Bloomington (2003), available at https://cogsci.indiana.edu/pub/hof.just-who-will-be-we.pdf. The

longer story of Hofstadter's disenfranchisement is captured in a wonderful piece: James Somers, "The Man Who Would Teach Machines to Think," *Atlantic*, November 2013.

6. The full quotation is from Gustavo Feigenbaum, *Conversations with John Searle* (Buenos Aires: Libros En Red, 2003), p. 57: "The only exceptions are things like Deep Blue, where you get enormous computational power, but where you're no longer trying to do A.I. at all. You're not trying to imitate humans on processing levels," and p. 58: "In a way, Deep Blue is giving up on A.I. because it doesn't say, 'Well we're going to try to do what human beings do,' but it says 'We're just going to overpower them with brute force.'"
7. Garry Kasparov, "The Chess Master and the Computer," *New York Review of Books*, 11 February 2010.
8. Quoted in William Herkewitz, "Why Watson and Siri Are Not Real AI," *Popular Mechanics*, 10 February 2014.
9. John Searle, "Watson Doesn't Know It Won on 'Jeopardy!'," *Wall Street Journal*, 23 February 2011.
10. Douglas Hofstadter, *Gödel, Escher, Bach: An Eternal Golden Braid* (London: Penguin, 2000), p. 601: "There is a related 'Theorem' about progress in AI: once some mental function is programmed, people soon cease to consider it as an essential ingredient of 'real thinking.' The ineluctable core of intelligence is always in that next thing which hasn't yet been programmed. This 'Theorem' was first proposed to me by Larry Tesler, so I call it *Tesler's Theorem*: 'AI is whatever hasn't been done yet.'"
11. Hofstadter, *Gödel, Escher, Bach*, p. 678.
12. Douglas Hofstadter, "Staring Emmy Straight in the Eye—And Doing My Best Not to Flinch" in David Cope, ed., *Virtual Music: Computer Synthesis of Musical Style* (London: MIT Press, 2004), p. 34.
13. Weber, "Mean Chess-Playing Computer Tears at Meaning of Thought."
14. Years later, though, he would wittily concede having "had to eat humble pie" with respect to his chess prediction and admit his mistake; Hofstadter, "Staring Emmy," p. 35.
15. Garry Kasparov, *Deep Thinking* (London: John Murray, 2017), pp. 251–52.
16. Quoted in Brad Leithhauser, "Kasparov Beats Deep Thought," *New York Times*, 14 January 1990.
17. Kasparov, "The Chess Master and the Computer."
18. See Dennett, *From Bacteria to Bach and Back*, p. 36.
19. Charles Darwin, *On the Origin of Species* (London: Penguin Books, 2009), p. 427.
20. See Isaiah Berlin, *The Hedgehog and the Fox* (New York: Simon & Schuster, 1953).
21. The distinction between AGI and ANI is often conflated with another one made by John Searle, who speaks of the difference between "strong" AI and "weak" AI. But the two are not the same thing at all. AGI and ANI reflect the breadth of a machine's capability, while Searle's terms describe whether a machine thinks like a human being ("strong") or unlike one ("weak").
22. Nick Bostrom and Eliezer Yudkowsky, "The Ethics of Artificial Intelligence" in William Ramsey and Keith Frankish, eds., *Cambridge Handbook of Artificial Intelligence* (Cambridge: Cambridge University Press, 2011).
23. Irving John Good, "Speculations Concerning the First Ultraintelligent Machine," *Advances in Computers* 6 (1966): 31–88.
24. Rory Cellan-Jones, "Stephen Hawking Warns Artificial Intelligence Could End Mankind," *BBC News*, 2 December 2014; Samuel Gibbs, "Elon Musk: AI 'Vastly More

Risky Than North Korea,'" *Guardian*, 14 August 2017; Kevin Rawlinson, "Microsoft's Bill Gates Insists AI Is a Threat," BBC News, 29 January 2015.

25. See Nick Bostrom, "Ethical Issues in Advanced Artificial Intelligence" in George Lasker, Wendell Wallach, and Iva Smit, eds., *Cognitive, Emotive, and Ethical Aspects of Decision Making in Humans and in Artificial Intelligence* (Windsor, ON: International Institute of Advanced Studies in Systems Research and Cybernetics, 2003), pp. 12–17.
26. Tad Friend, "How Frightened Should We Be of AI," *New Yorker*, 14 May 2018.
27. Volodymyr Mnih, Koray Kavukcuoglu, David Silver, et al., "Human-Level Control Through Deep Reinforcement Learning," *Nature* 518 (25 February 2015): 529–33.
28. David Autor, Frank Levy, and Richard Murnane, "The Skill Content of Recent Technological Change: An Empirical Exploration," *Quarterly Journal of Economics* 118, no. 4 (2003): 129–333. Another "non-routine" task listed was "forming/testing hypotheses." AlphaFold, a system developed by DeepMind to predict the 3-D structure of proteins, is a good example of progress made in this domain as well.
29. See ibid.; Autor and Dorn, "The Growth of Low-Skill Service Jobs"; and Autor, "Why Are There Still So Many Jobs?" I first developed this argument in Susskind, "Technology and Employment." This section draws in particular on my article "Re-Thinking the Capabilities of Technology in Economics," *Economics Bulletin* 39, no. 1 (2019): A30.
30. See Autor, "Polanyi's Paradox and the Shape of Employment Growth," p. 130; and Dana Remus and Frank Levy, "Can Robots Be Lawyers? Computers, Lawyers, and the Practice of Law," *Georgetown Journal of Legal Ethics* 30, no. 3 (2017): 501–58, available at https://papers.ssrn.com/sol3/papers.cfm?abstract_id=2701092.
31. See Demis Hassabis, "Artificial Intelligence: Chess Match of the Century," *Nature* 544 (2017): 413–14.
32. Cade Metz, "How Google's AI Viewed the Move No Human Could Understand," *Wired*, 14 March 2016. Also see Max Tegmark, *Life 3.0: Being Human in the Age of Artificial Intelligence* (London: Penguin Books, 2017), p. 87.
33. Cade Metz, "The Sadness and Beauty of Watching Google's AI Play Go," *Wired*, 3 November 2016.
34. "Beautiful," quoted in the preface to the paperback edition of Susskind and Susskind, *Future of the Professions*; "physically unwell" is from "Don't Forget Humans Created the Computer Program That Can Beat Humans at Go," *FiveThirtyEight*, 16 March 2016, https://fivethirtyeight.com/features/dont-forget-humans-created-the-computer-program-that-can-beat-humans-at-go/.
35. See, for instance, the Explainable AI Program at DARPA.
36. See Andrew Selbst and Julia Powles, "Meaningful Information and the Right to Explanation," *International Data Privacy Law* 7, no. 4 (2017): 233–42.
37. Quoted from Susskind and Susskind, *Future of the Professions*, p. 45.
38. Martin Marshall, "No App or Algorithm Can Ever Replace a GP, Say RCGP," 27 June 2018, available at https://www.gponline.com/ (accessed August 2018).
39. Daniel Susskind and Richard Susskind, "The Future of the Professions," *Proceedings of the American Philosophical Society* (2018). It is called "parametric design" since a family of possible buildings or objects are modeled through a set of adjustable "parameters" or variables. When the parameters are tweaked, the model generates a new version. See Susskind and Susskind, *Future of the Professions*, p. 95.

40. George Johnson, "Undiscovered Bach? No, a Computer Wrote It," *New York Times*, 11 November 1997.
41. Hofstadter, "Staring Emmy," p. 34.
42. Hofstadter, *Gödel, Escher, Bach*, p. 677.
43. H. A. Shapiro, "'Heros Theos': The Death and Apotheosis of Herakles," *Classical World* 77, no 1 (1983): 7–18. Quotation is from Hesiod, *Theogony, Works and Days, Testimonia*, Glenn W. Most, ed. and trans., Loeb Classical Library 57 (London: Harvard University Press, 2006), lines 950–55 of "Theogony."
44. Daniel Dennett calls this "cosmic warehouse" the "design space." See, for instance, Dennett, *From Bacteria to Bach and Back*.
45. This point was also made well by Sam Harris, the US neuroscientist, in his TED talk, "Can We Build AI Without Losing Control over It?," 29 September 2016.

5. TASK ENCROACHMENT

1. David Deming, "The Growing Importance of Social Skills in the Labor Market," *Quarterly Journal of Economics* 132, no. 4 (2017): 1593–1640.
2. Aaron Smith and Janna Anderson, "AI, Robotics, and the Future of Jobs: Key Findings," Pew Research Center, 6 August 2014, available at http://www.pewinternet.org/2014/08/06/future-of-jobs/ (accessed August 2018).
3. See, for instance, Erik Brynjolfsson and Tom Mitchell, "What Can Machine Learning Do? Workforce Implications," *Science* 358, no 6370 (2017).
4. John Markoff, "How Many Computers to Identify a Cat? 16,000," *New York Times*, 25 June 2012.
5. Jeff Yang, "Internet Cats Will Never Die," CNN, 2 April 2015.
6. Colin Caines, Florian Hoffman, and Gueorgui Kambourov, "Complex-Task Biased Technological Change and the Labor Market," International Finance Division Discussion Papers 1192 (2017).
7. Andrew Ng, "What Artificial Intelligence Can and Can't Do Right Now," *Harvard Business Review*, 9 November 2016.
8. "After forty years one begins to be able to distinguish an ephemeral surface ripple from a deeper current or an authentic change." From Antonia Weiss, "Harold Bloom, the Art of Criticism No. 1," *Paris Review* 118 (Spring 1991).
9. The consequences of "the encroachment of increasingly capable machines on those tasks that, until very recently, were considered necessarily human ones" is explored in Susskind, "Technology and Employment"; and "A Model of Technological Unemployment," Oxford University Department of Economics Discussion Paper Series No. 819 (2017).
10. Daniel Bell, "The Bogey of Automation," *New York Review of Books*, 26 August 1965.
11. For tractors, see Spencer Feingold, "Field of Machines: Researchers Grow Crop Using Only Automation," CNN, 8 October 2017; for milking, see Tom Heyden, "The Cows That Queue Up to Milk Themselves," BBC News, 7 May 2015; for herding, see Heather Brady, "Watch a Drone 'Herd' Cattle Across Open Fields," *National Geographic*, 15 August 2017; for cotton, see Virginia Postrel (2018), "Lessons from a Slow-Motion Robot Takeover," *Bloomberg View*, 9 February 2018.
12. For apples, see Tom Simonite, "Apple-Picking Robot Prepares to Compete for Farm Jobs," *MIT Technology Review*, 3 May 2017; for oranges, see Eduardo Porter, "In Florida

Groves, Cheap Labor Means Machines," *New York Times*, 22 March 2004; for grapes, see http://wall-ye.com/.

13. For wearables, see Khalil Akhtar, "Animal Wearables, Robotic Milking Machines Help Farmers Care for Cows," CBC News, 2 February 2016; for camera systems, see Black Swift Technologies Press Report, "Black Swift Technologies and NASA Partner to Push Agricultural Drone Technology Beyond NDVI and NDRE (Red Edge)," 20 March 2018; for autonomous sprayers, see James Vincent, "John Deere Is Buying an AI Startup to Help Teach Its Tractors How to Farm," *Verge*, 7 September 2017.
14. Jamie Susskind, *Future Politics* (Oxford: Oxford University Press, 2018), p. 54.
15. See Feingold, "Field of Machines."
16. Sidney Fussell, "Finally, Facial Recognition for Cows Is Here," *Gizmodo*, 1 February 2018.
17. James Vincent, "Chinese Farmers Are Using AI to Help Rear the World's Biggest Pig Population," *Verge*, 16 February 2018.
18. See, for instance, Adam Grzywaczewski, "Training AI for Self-Driving Vehicles: The Challenge of Scale," *NVIDIA Developer Blog*, 9 October 2017.
19. See their statement at https://corporate.ford.com/innovation/autonomous-2021.html (accessed 1 May 2018).
20. "All Tesla Cars Being Produced Now Have Full Self-Driving Hardware," 19 October 2019, https://www.tesla.com/en_GB/blog/all-tesla-cars-being-produced-now-have-full-self-driving-hardware (accessed 23 July 2019).
21. There are 1.25 million road fatalities a year, and between 20 and 50 million road injuries a year. See http://www.who.int/en/news-room/fact-sheets/detail/road-traffic-injuries (accessed 27 April 2018).
22. Joon Ian Wong, "A Fleet of Trucks Just Drove Themselves Across Europe," *Quartz*, 6 April 2016.
23. Sam Levin, "Amazon Patents Beehive-Like Structure to House Delivery Drones in Cities," *Guardian*, 26 June 2017; Arjun Kharpal, "Amazon Wins Patent for a Flying Warehouse That Will Deploy Drones to Deliver Parcels in Minutes," *CNBC*, 30 December 2016.
24. Nick Wingfield, "As Amazon Pushes Forward with Robots, Workers Find New Roles," *New York Times*, 10 September 2017.
25. For "harsh terrain," see "Cable-Laying Drone Wires Up Remote Welsh Village," BBC News, 30 November 2017; for "knotting ropes," see Susskind and and Susskind, *Future of the Professions*, p. 99; for "backflip," see Matt Simon, "Boston Dynamics" Atlas Robot Does Backflips Now and It's Full-Title Insane," *Wired*, 16 November 2017; and for others, see J. Susskind, *Future Politics*, p. 54.
26. Data in "Robots Double Worldwide by 2020: 3 Million Industrial Robots Use by 2020," *International Federation of Robotics*, 30 May 2018, https://ifr.org/ifr-press-releases/news/robots-double-worldwide-by-2020 (accessed August 2018). 2017 data from Statista, https://www.statista.com/statistics/947017/industrial-robots-global-operational-stock/ (accessed April 2019).
27. Ibid.
28. J. Susskind, *Future Politics*, p. 54.
29. Michael Chui, Katy George, James Manyika, and Mehdi Miremadi, "Human + Machine: A New Era of Automation in Manufacturing," McKinsey & Co., September 2017.
30. Carl Wilkinson, "Bot the Builder: The Robot That Will Replace Bricklayers," *Financial Times*, 23 February 2018.

31. Evan Ackerman, "AI Startup Using Robots and Lidar to Boost Productivity on Construction Sites," *IEEE Spectrum*, 24 January 2018.
32. See https://www.balfourbeatty.com/innovation2050 (accessed April 2019).
33. Alan Burdick, "The Marriage-Saving Robot That Can Assemble Ikea Furniture, Sort Of," *New Yorker*, 18 April 2018.
34. For "kippot," see Eitan Arom, "The Newest Frontier in Judaica: 3D Printing Kippot," *Jerusalem Post*, 24 October 2014; for all others, see J. Susskind, *Future Politics*, pp. 56–7.
35. Tomas Kellner, "Mind Meld: How GE and a 3D-Printing Visionary Joined Forces," *GE Reports*, 10 July 201710; "3D Printing Prosthetic Limbs for Refugees," *Economist*, 18 January 2018, https://www.youtube.com/watch?v=_W1veGQxMe4 (accessed August 2018).
36. Debra Cassens Weiss, "JPMorgan Chase Uses Tech to Save 460,000 Hours of Annual Work by Lawyers and Loan Officers," *ABA Journal*, 2 March 2017.
37. "Allen & Overy and Deloitte Tackle OTC Derivatives Market Challenge," 13 June 2016, http://www.allenovery.com/news/en-gb/articles/Pages/AllenOvery-and-Deloitte-tackle-OTC-derivatives-market-challenge.aspx (accessed August 2018).
38. Daniel Marin Katz, Michael J. Bommarito II, and Josh Blackman, "A General Approach for Predicting the Behavior of the Supreme Court of the United States," *PLOS ONE*, 12 April 2017, and Theodore W. Ruger, Pauline T. Kim, Andrew D. Martin, and Kevin M. Quinn, "The Supreme Court Forecasting Project: Legal and Political Science Approaches to Predicting Supreme Court Decisionmaking," *Columbia Law Review* 104:4 (2004), 1150–1210.
39. Nikolas Aletras, Dimitrios Tsarapatsanis, Daniel Preoţiuc-Pietro, and Vasileios Lampos, "Predicting Judicial Decisions of the European Court of Human Rights: A Natural Language Processing Perspective," *PeerJ Computer Science* 2:93 (2016).
40. Though by no means limited to diagnosis. See Eric Topol, "High-Performance Medicine: The Convergence of Human and Artificial Intelligence," *Nature* 25 (2019), 44–56, for a broader overview of the uses of AI in medicine.
41. Jeffrey De Fauw, Joseph Ledsam, Bernardino Romera-Paredes, et al., "Clinically Applicable Deep Learning for Diagnosis and Referral in Retinal Disease," *Nature Medicine* 24 (2018), 1342–50.
42. Pallab Ghosh, "AI Early Diagnosis Could Save Heart and Cancer Patients," *BBC News*, 2 January 2018.
43. Echo Huang, "A Chinese Hospital Is Betting Big on Artificial Intelligence to Treat Patients," *Quartz*, 4 April 2018.
44. Susskind and Susskind, *Future of the Professions*, p. 48.
45. Ibid., p. 58.
46. These statistics are from ibid., pp. 57–58.
47. Ibid., p. 56; Adam Thomson, "Personalised Learning Starts to Change Teaching Methods," *Financial Times*, 5 February 2018.
48. Steven Pearlstein, "The Robots-vs.-Robots Trading That Has Hijacked the Stock Market," *Washington Post*, 7 February 2018.
49. "Japanese Insurance Firm Replaces 34 Staff with AI," *BBC News*, 5 January 2017.
50. Hedi Ledford, "Artificial Intelligence Identifies Plant Species for Science," *Nature*, 11 August 2017; Jose Carranza-Rojas, Herve Goeau and Pierre Bonnet, "Going Deeper in the Automated Identification of Herbarium Specimens," *BMC Evolutionary Biology*

17:181 (2017). The system was first trained on about one million generic images from ImageNet, and then retrained on herbarium sheets.

51. Susskind and Susskind, *Future of the Professions*, p. 77; Jaclyn Peiser, "The Rise of the Robot Reporter," *New York Times*, 5 February 2019.
52. Cathy O'Neil, *Weapons of Math Destruction: How Big Data Increases Inequality and Threatens Democracy* (New York: Crown, 2016), p. 114, quoted in J. Susskind, *Future Politics*, p. 266.
53. Ibid., p. 31.
54. The Literary Creative Turing Tests are hosted by the Neukom Institute for Computational Science at Dartmouth College; see http://bregman.dartmouth.edu/turingtests/ (accessed August 2018).
55. See, for instance, Simon Colton and Geraint Wiggins, "Computational Creativity: The Final Frontier?" *Proceedings of the 20th European Conference on Artificial Intelligence* (2012), 21–6.
56. See "UN to Host Talks on Use of 'Killer Robot,'" *VOA News*, Agence France-Presse, 10 November 2017.
57. See, for instance, Joshua Rothman, "In the Age of AI, Is Seeing Still Believing?," *New Yorker*, 12 November 2018.
58. Javier C. Herńandez, "China's High-Tech Tool to Fight Toilet Paper Bandits," *New York Times*, 20 March 2017.
59. Dan Gilgoff and Hada Messia, "Vatican Warns About iPhone Confession App," *CNN*, 10 February 2011.
60. J. Susskind, *Future Politics*, p. 52.
61. See Ananya Bhattacharya, "A Chinese Professor Is Using Facial Recognition to Gauge How Bored His Students Are," *Quartz*, 12 September 2016.
62. For "woman and a child" see Raffi Khatchadourian, "We Know How You Feel," *New Yorker*, 19 January 2015. For "way a person walks," see J. Susskind, *Future Politics*, p. 53.
63. Khatchadourian, "We Know How You Feel," and Zhe Wu et al., "Deception Detection in Videos," https://arxiv.org/, 12 December 2017.
64. Alexandra Suich Bass, "Non-tech Businesses Are Beginning to Use Artificial Intelligence at Scale," *Economist*, 31 March 2018.
65. J. Susskind, *Future Politics*, p. 54.
66. http://www.parorobots.com/; *BBC News*, "Pepper Robot to Work in Belgian Hospitals," 14 June 2016.
67. Marc Ambasna-Jones, "How Social Robots Are Dispelling Myths and Caring for Humans," *Guardian*, 9 May 2016.
68. http://khanacademyannualreport.org/.
69. See Norri Kageki, "An Uncanny Mind: Masahiro Mori on the Uncanny Valley and Beyond," *IEEE Spectrum*, 12 June 2012.
70. Olivia Solon, "The Rise of 'Pseudo-AI': How Tech Firms Quietly Use Humans to Do Bots' Work," *Guardian*, 6 July 2018.
71. Aliya Ram, "Europe's AI Start-Ups Often Do Not Use AI, Study Finds," *Financial Times*, 5 March 2019.
72. Matthew DeBord, "Tesla's Future Is Completely Inhuman—and We Shouldn't Be Surprised," *Business Insider UK*, 20 May 2017; Kirsten Korosec, 1 November 2017, https://twitter.com/kirstenkorosec/status/925856398407213058.

73. https://twitter.com/elonmusk/status/984882630947753984 (accessed April 2019).
74. Data is from a search of the CB Insights database of earnings calls at https://www.cbinsights.com/. Similar searches are done in "On Earnings Calls, Big Data Is Out. Execs Have AI on the Brain," *CB Insights*, 30 November 2017; Bass, "Non-tech Businesses."
75. Paul Krugman, "Paul Krugman Reviews 'The Rise and Fall of American Growth' by Robert J. Gordon," *New York Times*, 25 January 2016.
76. Robert Gordon, *The Rise and Fall of American Growth* (Oxford: Princeton University Press, 2017).
77. In eighty-seven years' time because $100 \times 1.008^{87} = 200.01$, to two decimal places. If the United States were to return to the 2.41 percent growth rate, the same doubling of wealth would take just twenty-nine years: $100 \times 1.0241^{29} = 199.50$. Thomas Piketty makes a similar point in *Capital in the Twenty-First Century* (London: Harvard University Press, 2014), p. 5, noting that "the right way to look at the problem is once again in generational terms. Over a period of thirty years, a growth rate of 1 percent per year corresponds to cumulative growth of more than 35 percent. A growth rate of 1.5 percent per year corresponds to cumulative growth of more than 50 percent. In practice, this implies major changes in lifestyle and employment."
78. Gordon, Robert, *The Rise and Fall of American Growth* (Oxford: Princeton University Press, 2017), p. 96.
79. "A Study Finds Nearly Half of Jobs Are Vulnerable to Automation," *Economist*, 24 April 2018.
80. Carl Frey, Michael Osborne, Craig Holmes, et al., "Technology at Work v2.0: The Future Is Not What It Used to Be," Oxford Martin School and Citi (2016).
81. Automation risk data from Ljudiba Nedelkoska and Glenda Quintini, "Automation, Skills Use and Training," OECD Social, Employment and Migration Working Papers, No. 202 (2018); GDP per person data is OECD data for 2016 (retrieved 2018). PPPs (purchasing power parities) are currency exchange rates that try to take account of different price levels across different countries.
82. OECD, *Job Creation and Local Economic Development 2018: Preparing for the Future of Work* (Paris: OECD Publishing, 2018), p. 26. These comparisons use the same measure of "automation risk": if it has a 70 percent or higher probability of being automated (see p. 42).
83. I faintly remember the economist Robert Allen telling this anecdote in lectures when I was a graduate student. I am grateful to him for the tale.
84. Jonathan Cribb, Robert Joyce, and Agnes Norris Keiller, "Will the Rising Minimum Wage Lead to More Low-Paid Jobs Being Automated?," Institute for Fiscal Studies, 4 January 2018.
85. GreenFlag, "Automatic Car Washes Dying Out as the Hand Car Wash Cleans Up," published online at http://blog.greenflag.com/2015/automatic-car-washes-dying-out-as-the-hand-car-wash-cleans-up/ (accessed September 2018).
86. Robert Allen, "Why Was the Industrial Revolution British?," *Voxeu*, 15 May 2009.
87. Leo Lewis, "Can Robots Make Up for Japan's Care Home Shortfall?," *Financial Times*, 18 October 2017. The antipathy toward foreign workers may be changing, though. For instance, see "Japan Is Finally Starting to Admit More Foreign Workers," *Economist*, 5 July 2018.

88. Lewis, "Can Robots Make Up" and Joseph Quinlan, "Investors Should Wake Up to Japan's Robotic Future," *Financial Times*, 25 September 2017.
89. Daron Acemoglu and Pascual Restrepo, "Demographics and Automation," NBER Working Paper No. 24421 (2018).
90. From the Data Appendix to William Nordhaus, "Two Centuries of Productivity Growth in Computing," *Journal of Economic History* 67:1 (2007), 128–59. I thank William Nordhaus for sharing his revised data with me.
91. Quoted in Susskind and Susskind, *Future of the Professions*, p. 157.
92. Tom Simonite, "For Superpowers, Artificial Intelligence Fuels New Global Arms Race," *Wired*, 8 September 2017; "Premier Li Promotes Future of AI as Economic Driver," State Council, People's Republic of China, 23 July 2017, http://english.gov.cn/premier/news/2017/07/24/content_281475750043336.htm (accessed September 2018).
93. Aron Smith, "Public Attitudes Toward Computer Algorithms," Pew Research Center, November 2018.
94. Daisuke Wakabayashi and Cade Metz, "Google Promises Its A.I. Will Not Be Used for Weapons," *New York Times*, 7 June 2018; Hal Hodson, "Revealed: Google AI Has Access to Huge Haul of NHS Patient Data," *New Scientist*, 29 April 2016, and the response from DeepMind, https://deepmind.com/blog/ico-royal-free/ (accessed August 2018).
95. Eric Topol, "Medicine Needs Frugal Innovation," *MIT Technology Review*, 12 December 2011.
96. Frey, "Technology at Work v2.0."
97. Steve Johnson, "Chinese Wages Now Higher Than in Brazil, Argentina and Mexico," *Financial Times*, 26 February 2017.
98. Ben Bland, "China's Robot Revolution," *Financial Times*, 6 June 2016.
99. "China's Robot Revolution May Affect the Global Economy," *Bloomberg News*, 22 August 2017.
100. Michael Wooldridge, "China Challenges the US for Artificial Intelligence Dominance," *Financial Times*, 15 March 2018.
101. "Tsinghua University May Soon Top the World League in Science Research," *The Economist*, 17 November 2018.

6. FRICTIONAL TECHNOLOGICAL UNEMPLOYMENT

1. Keynes, *Essays in Persuasion*, p. 364.
2. Homer, *Odyssey*, book XI.
3. Economists often make a distinction between "structural" unemployment and "frictional" unemployment. As far as I can see, this distinction between the two types of *technological* unemployment is a new one.
4. This is men between the ages of twenty and sixty-four; Nicholas Eberstadt, *Men Without Work: America's Invisible Crisis* (West Conshohocken, PA: Templeton Press, 2016).
5. YiLi Chien and Paul Morris, "Is U.S. Manufacturing Really Declining?," *Federal Bank of St. Louis Blog*, 11 April 2017 (accessed 23 July 2019).
6. US GDP per capita growth averaged about 2 percent per annum. See Charles I. Jones, "The Facts of Economic Growth" in John B. Taylor and Harald Uhlig, eds., *Handbook of Macroeconomics*, vol. 2A (Amsterdam: Elsevier, 2016), pp. 3–69.

7. David Autor, "Work of the Past, Work of the Future," Richard T. Ely Lecture delivered at the Annual Meeting of the American Economic Association (2019), from 7:40 in https://www.aeaweb.org/webcasts/2019/aea-ely-lecture-work-of-the-past-work-of-the-future (accessed January 2019).
8. Ryan Avent, *The Wealth of Humans: Work and Its Absence in the 21st Century* (London: Allen Lane, 2016), p. 53.
9. See, for instance, Claudia Goldin and Lawrence Katz, *The Race Between Education and Technology* (London: Harvard University Press, 2009).
10. Avent, *The Wealth of Humans*, p. 55.
11. Stuart W. Elliott, "Computers and the Future of Skill Demand," *OECD Educational Research and Innovation* (2017), p. 96.
12. This is known as the "postgraduate wage premium." See Joanne Lindley and Stephen Machin, "The Rising Postgraduate Wage Premium," *Economica* 83 (2016): 281–306, and also figure 6 in Autor, "Skills, Education, and the Rise of Earnings Inequality."
13. Glenn Thrust, Nick Wingfield, and Vindu Goel, "Trump Signs Order That Could Lead to Curbs on Foreign Workers," *New York Times*, 18 April 2017.
14. See, for instance, Norman Matloff, "Silicon Valley Is Using H-1B Visas to Pay Low Wages to Foreign Workers," *Medium*, 23 March 2018. (Matloff is a professor of computer science at UC Davis.)
15. Jean-François Gagné, "Global AI Talent Report 2018," http://www.jfgagne.ai/talent (accessed August 2018). Given that these estimates are derived from LinkedIn data, a Western platform, they are likely to underestimate the total number, and so overestimate the proportion working in the United States.
16. Autor, "Work of the Past, Work of the Future."
17. Annie Lowrey, *Give People Money: The Simple Idea to Solve Inequality and Revolutionise Our Lives* (London: W. H. Allen, 2018), p. 37.
18. Edward Luce, *The Retreat of Western Liberalism* (London: Little, Brown, 2017), p. 53.
19. Paul Beaudry, David Green, and Benjamin Sand, "The Great Reversal in the Demand for Skill and Cognitive Tasks," *Journal of Labor Economics* 34:1 (2016): 199–247, quoted in "Special Report on Lifelong Education: Learning and Earning," *Economist*, 14 January 2017, p. 2.
20. "Time to End the Academic Arms Race," *Economist*, 3 February 2018.
21. Chang May Choon, "Dream Jobs Prove Elusive for South Korea's College Grads," *Straits Times*, 11 March 2016.
22. For "pink-collar" see, for instance, Elise Kalokerinos, Kathleen Kjelsaas, Steven Bennetts, and Courtney von Hippel, "Men in Pink Collars: Stereotype Threat and Disengagement Among Teachers and Child Protection Workers," *European Journal of Social Psychology* 47:5 (2017); for the percentages, see the US Bureau of Labor Statistics, "Household Data: Annual Averages" for 2017 at https://www.bls.gov/cps/cpsaat11.pdf (accessed August 2018).
23. Gregor Aisch and Robert Gebeloff, "The Changing Nature of Middle-Class Jobs," *New York Times*, 22 February 2015. The Bureau of Labor Statistics data for 2017 again shows the domination of manufacturing roles by men.
24. As Lawrence Katz puts it, this is not a problem of "skills-mismatch," but "identity-mismatch." Claire Cain Miller, "Why Men Don't Want the Jobs Done Mostly by Women," *New York Times*, 4 January 2017.

25. US Bureau of Labor Statistics, "Protections of Occupational Employment, 2014–24," *Career Outlook*, December 2015. In terms of low pay, the exception is nursing—most OECD countries pay their nurses above the national average, though some, like the UK, only marginally above, and others, like France, pay below. See Exhibit 13 in Adair Turner (Institute for New Economic Thinking), "Capitalism in an Age of Robots," presentation at the School of Advanced International Studies, Washington, DC, 10 April 2018, https://www.ineteconomics.org/uploads/papers/Slides-Turner-Capitalism-in-the-Age-of-Robots.pdf (accessed August 2018). Also see OECD, "Health at a Glance 2017: OECD Indicators" (February 2018), chap. 8, p. 162, on "remuneration of nurses." In 2015, the United States paid its nurses 1.24 times the average wage; the UK, 1.04; and France, 0.95.
26. Personal care aides (83.7 percent), registered nurses (89.9 percent), home health aides (88.6 percent), food preparation and services (53.8 percent), retail salespersons (48.2 percent). Again, see the US Bureau of Labor Statistics "Household Data."
27. Enrico Moretti, *The New Geography of Jobs* (New York: First Mariner Books, 2013), p. 17.
28. Ibid., p. 23.
29. Ibid., pp. 82–5.
30. Ibid., p. 89.
31. Emily Badger and Quoctrung Bui, "What If Cities Are No Longer the Land of Opportunity for Low-Skilled Workers?," *New York Times*, 11 January 2019.
32. Moretti, *New Geography of Jobs*.
33. Eurostat (2019) data, https://ec.europa.eu/eurostat/statistics-explained/index.php?title=Young_people_-_social_inclusion#Living_with_parents (accessed April 2019).
34. Moretti, *New Geography of Jobs*, p. 157.
35. Louis Uchitelle, "Unemployment Is Low, but That's Only Part of the Story," *New York Times*, 11 July 2019.
36. Consider, for instance, Benjamin Friedman, a Harvard economist, writing that "the question is not whether millions of would-be-workers will be chronically out of work . . . most Americans will find something to do. But far too many of the jobs they will end up taking will pay them too little support what our society considers a middle-class standard of living." Benjamin M. Friedman, "Born to Be Free," *New York Review of Books*, 12 October 2017.
37. From Bureau of Labor Statistics, "Profile of the Working Poor, 2016," https://www.bls.gov/opub/reports/working-poor/2016/home.htm (accessed July 2018).
38. Robert Reich, a public-policy professor and former secretary of labor for Bill Clinton, once estimated that by 2020 up to 40 percent of Americans would have "uncertain" work like this, the sort of work that makes up the "gig," "share," "irregular," or "precarious" economy, and by 2025, most workers will. This is likely to turn out to be an overestimate, though; in 2017, only 10 percent worked in so-called alternative work arrangements, a slight decline from 2005. Robert Reich, "The Sharing Economy Will Be Our Undoing," *Salon*, 25 August 2015; Ben Casselman, "Maybe the Gig Economy Isn't Reshaping Work After All," *New York Times*, 7 June 2018.
39. Andy Haldane, "Labour's Share," speech at the Trades Union Congress, London, 12 November 2015; Richard Partington, "More Regular Work Wanted by Almost Half Those on Zero-Hours," *Guardian*, 3 October 2018.

40. Quoted in Friedman, "Born to Be Free."
41. Tyler Cowen, *Average Is Over: Powering America Beyond the Age of the Great Stagnation* (New York: Dutton, 2013), p. 23.
42. Lowrey, *Give People Money*, p. 15.

7. STRUCTURAL TECHNOLOGICAL UNEMPLOYMENT

1. Chris Hughes, *Fair Shot: Rethinking Inequality and How We Earn* (London: Bloomsbury, 2018), p. 82.
2. The argument of this chapter runs through my doctorate, Susskind, "Technology and Employment." Parts of the argument can be found in my articles "A Model of Technological Unemployment" and "Automation and Demand," Oxford University Department of Economics Discussion Paper Series No. 845 (2018) as well.
3. But not completely irrelevant. Recall the discussion in chapter 5 about how both relative productivities *and* relative costs matter for deciding whether to automate a task: as with the case of the mechanical car wash, even if a machine is more productive than a worker, if that worker is willing to work for a lower wage than before, it may not make financial sense to use the machine.
4. I explore this example in "Robots Probably Won't Take Our Jobs—for Now," *Prospect*, 17 March 2017.
5. For instance, in Tyler Cowen's podcast, "Conversations with Tyler," episode 22 titled "Garry Kasparov on AI, Chess, and the Future of Creativity."
6. The new machine, dubbed AlphaZero, was matched up against the champion chess computer Stockfish. Of the fifty games where AlphaZero played white, it won twenty-five and drew twenty-five; of the fifty games where it played black, it won three and drew forty-seven. David Silver, Thomas Hubert, Julian Schrittwieser, et al., "Mastering Chess and Shogi by Self-Play with a General Reinforcement Learning Algorithm," https://arxiv.org, arXiv:1712.01815v1 (2017).
7. Tyler Cowen, "The Age of the Centaur Is Over Skynet Goes Live," *Marginal Revolution*, 7 December 2017.
8. See Kasparov, *Deep Thinking*, chap. 11.
9. Data is from Ryland Thomas and Nicholas Dimsdale, "A Millennium of UK Data," Bank of England OBRA data set (2017). The real GDP data is the spliced data from sheet A14; the employment data is the spliced data from A53. There are gaps in both series during the First and Second World Wars, and employment data is only available for the first year of each decade from 1861 through 1911; I have interpolated the data between those points in this figure. Real GDP data for 1861–71 is GB, not UK. https://www.bankofengland.co.uk/statistics/research-datasets.
10. "70 percent" and "30 percent" are calculated from Federal Reserve Bank of St. Louis (FRED) data; see https://fred.stlouisfed.org/tags/series?t=manufacturing (accessed October 2018); the 5.7 million is from Martin Baily and Barry Bosworth, "US Manufacturing: Understanding Its Past and Its Potential Future," *Journal of Economic Perspectives* 28:1 (2014): 3–26. As others have noted, US manufacturing itself may have declined as a share of nominal GDP over the last few decades, but not of real GDP; see, for instance, Chien and Morris, "Is U.S. Manufacturing Really Declining?"

11. Data is from Thomas and Dimsdale, "A Millennium of UK Data."
12. In Joel Mokyr, Chris Vickers, and Nicholas Ziebarth, "The History of Technological Anxiety and the Future of Economic Growth: Is This Time Different?" *Journal of Economic Perspectives* 29, no. 3 (2015): 31–50.
13. In David Autor and David Dorn, "Technology Anxiety Past and Present," Bureau for Employers' Activities, International Labour Office (2013).
14. Autor, "Polanyi's Paradox and the Shape of Employment Growth," p. 148.
15. Quoted in John Thornhill, "The Big Data Revolution Can Revive the Planned Economy," *Financial Times*, 4 September 2017.
16. Andre Tartar, "The Hiring Gap," *New York Magazine*, 17 April 2011; "Apple," https://www.forbes.com/companies/apple/; "Microsoft," https://www.forbes.com/companies/microsoft/ (accessed May 2019).
17. Edward Luce, *The Retreat of Western Liberalism* (London: Little, Brown, 2017), p. 54.
18. Thor Berger and Carl Frey, "Industrial Renewal in the 21st Century: Evidence from US Cities," *Regional Studies* (2015).
19. Data from Rodolfo Manuelli and Ananth Seshadri, "Frictionless Technology Diffusion: The Case of Tractors," *American Economic Review* 104, no. 4 (2014): 1268–1391.
20. See Daron Acemoglu and Pascual Restrepo, "The Race Between Machine and Man: Implications of Technology for Growth, Factor Shares, and Employment," *American Economic Review* 108, no. 6 (2018): 1488–542.
21. Wassily Leontief quoted in Nils Nilsson, "Artificial Intelligence, Employment, and Income," *AI Magazine*, Summer 1984. He shared similar reflections in Leonard Silk, "Economic Scene; Structural Joblessness," *New York Times*, 6 April 1983.
22. Acemoglu and Restrepo, "The Race Between Machine and Man." Horses and human beings differ in other ways (of course). Some economists point to the fact that people, unlike horses, can own machines, and so do not need to rely on work alone. People, unlike horses, can also vote—and might elect the "anti-tractor" party, opposed to whatever technology threatens their jobs.
23. This is one possible case in the model in Acemoglu and Restrepo, "The Race Between Machine and Man."
24. Acemoglu and Restrepo do not think it is *necessarily* the case that these new tasks will be created for human beings to do, however. For instance, in Daron Acemoglu and Pascual Restrepo, "The Wrong Kind of AI? Artificial Intelligence and the Future of Labor Demand," MIT Working Paper (2019), they explicitly consider the possibility that this does not happen.
25. In John Stuart Mill, *Principles of Political Economy with Some of Their Applications to Social Philosophy* (London: Longmans, Green, 1848), he states both that demand for commodities "does not constitute demand for labour" and, separately, that it "is not demand for labour." These are quoted in Susskind, "Technology and Employment."
26. Victor Mather, "Magnus Carlsen Wins World Chess Championship, Beating Fabiano Curuana," *New York Times*, 28 November 2018.
27. This is explored in Susskind and Susskind, *Future of the Professions*, pp. 244–45.
28. For economists using it, see "Automation and Anxiety," *Economist*; for technologists using it, see Marc Andreessen, "Robots Will Not Eat the Jobs But Will Unleash Our Creativity," *Financial Times*, 23 June 2014; for commentators, see Annie Lowrey,

"Hey, Robot: What Cat Is Cuter?," *New York Times Magazine*, 1 April 2014; for politicians, see Georgia Graham, "Robots Will Take Over Middle-Class Professions, Says Minister," *Telegraph*, 8 July 2014.

29. David Schloss, *Methods of Industrial Remuneration* (London: Williams and Norgate, 1898). The text has been archived online at https://ia902703.us.archive.org/30/items/methodsofindustr00schl/methodsofindustr00schl.pdf. The *Economist* website has an entry on the "lump of labour fallacy" and David Schloss at http://www.economist.com/economics-a-to-z/l. Tom Walker, an economist, has written at length about the idea and its origins, too; see, for instance, "Why Economists Dislike a Lump of Labor," *Review of Social Economy* 65, no. 3 (2007): 279–91.
30. Schloss, *Methods of Industrial Remuneration*, p. 81.
31. Leontief, "National Perspective," p. 4.
32. Daron Acemoglu and Pascual Restrepo, "Robots and Jobs: Evidence from US Labor Markets," NBER Working Paper No. 23285 (2017).
33. Quoted in Susan Ratcliffe, ed., *Oxford Essential Quotations*, 4th ed. (2016), at http://www.oxfordreference.com/ (accessed 13 May 2018).
34. That year, the Nazis won more seats in Parliament than any other party. The unemployment statistic is from Nicholas Dimsdale, Nicholas Horsewood, and Arthur Van Riel, "Unemployment in Interwar Germany: An Analysis of the Labor Market, 1927–1936," *Journal of Economic History* 66, no. 3 (2006): 778–808. The point itself came up in conversation with Tim Harford, the economist and journalist. I am grateful for his reflections.

8. TECHNOLOGY AND INEQUALITY

1. Jean-Jacques Rousseau, "The Genesis of Inequality," from *Discourse on the Origin and Foundation of Inequality Among Men* in David Johnston, *Equality* (Indianapolis: Hackett Publishing, 2000), chap. 5.
2. Walter Scheidel, *The Great Leveler: Violence and the History of Inequality from the Stone Age to the Twenty-First Century* (Oxford: Princeton University Press, 2017), p. 28.
3. Ibid., p. 33.
4. Harari, *Sapiens*, chap. 1.
5. Piketty, *Capital in the Twenty-First Century*, p. 48.
6. Arthur Pigou, *A Study in Public Finance* (London: Macmillan, 1928), p. 29; "Gary Becker's Concept of Human Capital," *Economist*, 3 August 2017.
7. Gary Becker, "The Economic Way of Looking at Life," Nobel Prize lecture, 9 December 1992.
8. George Orwell, *Essays* (London: Penguin Books, 2000), p. 151.
9. The story of this statistic is an ironic one. Though it is often used today as a measure of "fairness," its creator, Carrado Gini, was an enthusiastic fascist.
10. See both Era Dabla-Norris, Kalpana Kochhar, Frantisek Ricka, et al., "Causes and Consequences of Income Inequality: A Global Perspective," IMF Staff Discussion Note (2015); and Jan Luiten van Zanden, Joerg Baten, Marco Mira d'Ercole, et al., "How Was Life? Global Well-Being Since 1820," OECD (2014), p. 207: "It is hard not to notice the sharp increase in income inequality experienced by the vast majority of countries from the 1980s. There are very few exceptions to this . . ."

11. See, for instance, Piketty, *Capital in the Twenty-First Century*, p. 266.
12. These are post-tax and transfer Gini coefficients for 2017, or latest available year. This is an updated version of Figure 1.3 in OECD, "In It Together: Why Less Inequality Benefits All" (2015), using OECD (2019) data; http://www.oecd.org/social/income-distribution-database.htm (accessed April 2019).
13. John Rawls, *A Theory of Justice* (Cambridge, MA: Harvard University Press, 1999), p. 266.
14. These are pre-tax income, from Appendix Data FS40 in Thomas Piketty, Emmanuel Saez, and Gabriel Zucman, "Distribution National Accounts: Methods and Estimates for the United States," *Quarterly Journal of Economics* 133, no. 2 (2018): 553–609. The data is available at http://gabriel-zucman.eu/usdina/. The bottom 10 percent are omitted, as the authors note, since their pre-tax income is close to zero, and sometimes negative.
15. For the United States, see http://wid.world/country/usa/—it was 11.05 percent in 1981, 20.2 percent in 2014. For the UK, see http://wid.world/country/united-kingdom/—it was 6.57 percent in 1981 and 13.9 percent in 2014.
16. 0.1 percent: 2.23 percent in 1981, 7.89 in 2017. 0.01 percent: 0.66 percent in 1981, 3.44 percent in 2017 (excluding capital gains). Data is from Emmanuel Saez and Thomas Piketty, "Income Inequality in the United States, 1913–1998," *Quarterly Journal of Economics* 118, no. 1 (2003): 1–39, the "Table A1" in the online data appendix, at https://eml.berkeley.edu/~saez/ (accessed April 2019).
17. Jonathan Cribb, Andrew Hood, Robert Joyce, and Agnes Norris Keller, "Living Standards, Poverty and Inequality in the UK: 2017," Institute for Fiscal Studies, 19 July 2017.
18. Top 1 percent share of pre-tax national income. This is an update of figure 1 in OECD, "FOCUS on Top Incomes and Taxation in OECD Countries: Was the Crisis a Game Changer?" (May 2014), using the latest World Inequality Database data; https://wid.world/data/ (accessed April 2019).
19. In OECD countries, wages and salaries make up 75 percent of household incomes of working-age adults. See OECD, "Growing Income Inequality in OECD Countries: What Drives It and How Can Policy Tackle It?" (2011).
20. See OECD, "Promoting Productivity and Equality: A Twin Challenge," chap. 2 of *OECD Economic Outlook 2016*, vol. 1 (2016). On p. 69, the authors note how "inequality in the distribution of labour income has accounted for most of the increase in income inequality." Also see International Labour Organization, *Global Wage Report 2014/2015* (Geneva: International Labour Office, 2015), where they note that the OECD report "Divided We Stand: Why Inequality Keeps Rising" (2011) "documented how in developed economies, in the decades before the crisis, greater wage inequality had been the single most important driver of income inequality."
21. See OECD, "Growing Income Inequality in OECD Countries"; and Anthony B. Atkinson, *The Changing Distribution of Earnings in OECD Countries* (Oxford: Oxford University Press, 2009).
22. Emmanuel Saez, "Striking It Richer: The Evolution of Top Incomes in the United States," published online at https://eml.berkeley.edu/~saez/ (2016); Piketty, *Capital in the Twenty-First Century*, p. 315.
23. Emmanuel Saez and and Gabriel Zucman, "Wealth Inequality in the United States Since 1913: Evidence from Capitalized Income Tax Data," *Quarterly Journal of Economics* 131:2 (2016): 519–78. This is drawn from DataFig8-9b in the online data appendix, at http://gabriel-zucman.eu/.

24. Laura Tyson and Michael Spence, "Exploring the Effects of Technology on Income and Wealth Inequality" in Heather Boushey, J. Bradford DeLong, and Marshall Steinbaum, eds., *After Piketty: The Agenda for Economics and Inequality* (London: Harvard University Press, 2017), pp. 182–83.
25. Saez and Piketty, "Income Inequality in the United States." This figure is based on data from "Table B2" in the online data appendix, at https://eml.berkeley.edu/~saez/ (accessed April 2019).
26. From Lawrence Mishel and Alyssa Davis, "Top CEOs Make 300 Times More Than Typical Workers," *Economic Policy Institute*, 21 June 2015. In 1977, the ratio was 28.2, in 2000 it was 376.1, and in 2014 it had fallen to 303.4.
27. "It is rather remarkable how nearly constant are the proportions of the various categories over long periods of time, between both good years and bad. The size of the total social pie may wax and wane, but total wages seem always to add up to about two-thirds of the total." Paul Samuelson, quoted in Hagen Krämer, "Bowley's Law: The Diffusion of an Empirical Supposition into Economic Theory," *Papers in Political Economy* 61 (2011).
28. John Maynard Keynes, "Relative Movements of Real Wages and Output," *Economic Journal* 49, no. 93 (1939): 34–51; Nicholas Kaldor, "A Model of Economic Growth," *Economic Journal* 67, no. 268 (1957): 591–624; and Charles Cobb and Paul Douglas, "A Theory of Production," *American Economic Review* 18, no. 1 (1928): 139–65.
29. Loukas Karabarbounis and Brent Neiman, "The Global Decline of the Labor Share," *Quarterly Journal of Economics* 129, no. 1 (2014): 61–103.
30. Mai Chi Dao, Mitali Das, Zsoka Koczan, and Weicheng Lian, "Drivers of Declining Labor Share of Income," *IMF Blog* (2017).
31. Chapter 2 in OECD, *OECD Employment Outlook 2018* (Paris: OECD Publishing, 2018). Countries include Finland, Germany, Japan, Korea, the United States, France, Italy, Sweden, Austria, Belgium, UK, Australia, Spain, Czech Republic, Denmark, Hungary, Poland, Netherlands, Norway, Canada, New Zealand, Ireland, Israel, and the Slovak Republic. Here, "average wages" refers to "real median compensation."
32. Chi Dao et al., "Drivers of Declining Labor Share of Income."
33. It is chapter 3 of OECD, *OECD Employment Outlook 2012* (Paris: OECD Publishing, 2012), titled "Labour Losing to Capital: What Explains the Declining Labour Share?" that is quoted in part 1 of World Economic Forum's *Global Risks Report 2017*. The reality is subtler than the WEF's reading, though—technology, according to *OECD Employment Outlook 2012*, explains 80 percent of the "within-industry changes" in the labor share, and within-industry changes explain "an overwhelming proportion" of the aggregate fall in the labor share (rather than between-industry changes).
34. See both IMF, *World Economic Outlook* (2017), chap. 3; and Karabarbounis and Neiman, "Global Decline of the Labor Share."
35. Chi Dao et al., "Drivers of Declining Labor Share of Income."
36. Economic Policy Institute, "The Productivity-Pay Gap" (October 2017), available at http://www.epi.org/productivity-pay-gap/. Cumulative percent change since 1948 is on the *y*-axis.
37. David Autor, David Dorn, Lawrence Katz, et al., "The Fall of the Labor Share and the Rise of Superstar Firms," NBER Working Paper No. 23396 (2017).

38. Ibid.
39. Ibid.
40. PwC, "Global Top 100 Companies by Market Capitalisation" (2018). Alibaba and Amazon are nominally classified in the "consumer services" category, but both are better thought of as technology companies instead.
41. Piketty, *Capital in the Twenty-First Century*, p. 244.
42. Melanie Kramers, "Eight People Own Same Wealth as Half the World," Oxfam press release, 16 January 2017.
43. "Are Eight Men as Wealthy as Half the World's Population?," *Economist*, 19 January 2017.
44. Dabla-Norris, Kochhar, Ricka, et al., "Causes and Consequences of Income Inequality," p. 16.
45. Here, "richest" is "richest in wealth"; see Piketty, *Capital in the Twenty-First Century*, table 7.2, pp. 248–49 and 257.
46. Ibid., p. 257.
47. Joseph Stiglitz, "Inequality and Economic Growth," *Political Quarterly* 86, no. 1 (2016): 134–55.
48. Emmanuel Saez and Gabriel Zucman, "Wealth Inequality in the United States Since 1913: Evidence from Capitalized Income Tax Data," *Quarterly Journal of Economics* 131, no. 2 (2016): 519–78.
49. Thomas Piketty and Gabriel Zucman, "Capital Is Back: Wealth–Income Ratios in Rich Countries 1700–2010," *Quarterly Journal of Economics* 129, no. 3 (2014): 1255–1310.
50. Data is from DataFig1-6-7b in the online data appendix for Saez and Zucman, "Wealth Inequality in the United States." This chart is also in "Forget the 1%," *Economist*, 6 November 2014.
51. Facundo Alvaredo, Lucas Chancel, Thomas Piketty, et al., *World Inequality Report* (Creative Commons, 2018), p. 9.
52. Keynes, *Essays in Persuasion*, p. 360.
53. Ibid., p. 373.
54. Ibid., p. 367.
55. Joseph Stiglitz, "Towards a General Theory of Consumerism: Reflections on Keynes's Economic Possibilities for Our Grandchildren" in Lorenzo Pecchi and Gustavo Piga, eds., *Revisiting Keynes: Economics Possibilities for Our Grandchildren* (Cambridge, MA: MIT Press, 2008).
56. In my view, the economic problem is already shifting away from the traditional *growth* problem of making the pie bigger for everyone, and toward the *distribution* problem of making sure that everyone gets a decent slice. Other economists have made similar distinctions, though not always to the same end. David Autor in "Why Are There Still So Many Jobs?," for instance, distinguishes between the problems of "scarcity" and "distribution," but is skeptical we have solved the former. He asks:

    > Are we actually on the verge of throwing off the yoke of scarcity so that our primary economic challenge soon becomes one of distribution? Here, I recall the observations of economist, computer scientist, and Nobel laureate Herbert Simon (1966), who wrote at the time of the automation anxiety of the 1960s: "Insofar as they are economic problems at all, the world's problems in this generation and the next are problems of scarcity, not of intolerable abun-

dance. The bogeyman of automation consumes worrying capacity that should be saved for real problems . . ." A half century on, I believe the evidence favors Simon's view.

I disagree with Autor's conclusion, but find the framing useful.

57. See http://www.worldbank.org/en/topic/poverty/overview (accessed April 2018).

9. EDUCATION AND ITS LIMITS

1. See https://web.archive.org/web/20180115215736/twitter.com/jasonfurman/status/913439100165918721.
2. Enrico Moretti, *The New Geography of Jobs* (New York: First Mariner Books, 2013), p. 226.
3. Ibid., p. 228.
4. Claudia Goldin and Lawrence Katz, *The Race Between Education and Technology* (London: Harvard University Press, 2009), p. 13.
5. Ibid., p. 12.
6. Quoted in Michelle Asha Cooper, "College Access and Tax Credits," National Association of Student Financial and Administrators (2005).
7. Speech by Tony Blair, launching Labour's education manifesto at the University of Southampton, 23 May 2001, available at https://www.theguardian.com/politics/2001/may/23/labour.tonyblair.
8. Speech by President Barack Obama, "On Higher Education and the Economy," University of Texas at Austin, 9 August 2010.
9. "Special Report on Lifelong Education: Learning and Earning," *Economist*, 14 January 2017, p. 2.
10. Royal Society, *After the Reboot: Computing Education in UK Schools* (2017), pp. 52 and 53, respectively.
11. Ibid., p. 22. GCSEs are examinations typically taken by students aged 14–16 in England, Wales, and Northern Ireland.
12. "Special Report on Lifelong Education," p. 9.
13. Susskind and Susskind, *Future of the Professions*, p. 55.
14. Benjamin Bloom, "The 2 Sigma Problem: The Search for Methods of Group Instruction as Effective as One-to-One Tutoring," *Educational Researcher* 13, no. 6 (1984): 4–16. This is discussed ibid., p. 56.
15. See Susskind and Susskind, *Future of the Professions*, p. 58, n. 78.
16. Larry Summers made a similar case on Tyler Cowen's podcast, "Conversations with Tyler," episode 28, titled "Larry Summers on Macroeconomics, Mentorship, and Avoiding Complacency."
17. See, for instance, Seb Murray, "Moocs Struggle to Lift Rock-Bottom Completion Rates," *Financial Times*, 4 March 2019.
18. Joshua Goodman, Julia Melkers, and Amanda Pallais, "Can Online Delivery Increase Access to Education?" *Journal of Labor Economics* 37, no. 1 (2019).
19. Quoted in Tanja M. Laden, "Werner Herzog Hacks the Horrors of Connectivity in 'Lo and Behold,'" *Creators* on Vice.com, 25 August 2016.
20. Five hundred Singapore dollars, see http://www.skillsfuture.sg/credit.
21. Pew Research Center, "The State of American Jobs: The Value of a College Education,"

6 October 2016, http://www.pewsocialtrends.org/2016/10/06/5-the-value-of-a-college-education/ (accessed September 2018).

22. See "Tech Millionaire College Dropouts," *Guardian*, 11 January 2014; and "8 Inspiring Dropout Billionaires of the Tech Industry," *Times of India: Economic Times*, 11 April 2016.
23. "Thiel Fellows Skip or Stop Out of College," https://thielfellowship.org/ (accessed April 2019).
24. "Back to the Future with Peter Thiel," *National Review*, 20 January 2011.
25. Bryan Caplan, *The Case Against Education: Why the Education System Is a Waste of Time and Money* (Oxford: Princeton University Press, 2018), p. 4.
26. Gregory Ferenstein, "Thiel Fellows Program Is 'Most Misdirected Piece of Philanthropy,'" *TechCrunch*, 10 October 2013.
27. Avent, *The Wealth of Humans*, makes a similar argument to this. "University," he writes, "is hard. Many of those who don't currently make it through a college programme lack the cognitive ability to do so." See p. 55.
28. John F. Kennedy, "Moon Speech" at Rice Stadium, 12 September 1962: "We choose to go to the moon in this decade and do the other things, not because they are easy, but because they are hard," available at https://er.jsc.nasa.gov/seh/ricetalk.htm (accessed April 2019).
29. See Stuart W. Elliott, "Computers and the Future of Skill Demand," *OECD Educational Research and Innovation* (2017), p. 15.
30. See the abstract for ibid.
31. Yuval Harari, *Homo Deus: A Brief History of Tomorrow* (London: Harvill Secker, 2016), p. 269: "What will be the political impact of a massive new class of economically useless people?"; and from 61 minutes in "Yuval Harari with Dan Ariely: Future Think—From Sapiens to Homo Deus," published by 92nd Street Y on YouTube, 22 February 2017, https://www.youtube.com/watch?v=5BqD5klZsQE.

10. THE BIG STATE

1. See David Landes, *Abba Ptachya Lerner 1903–1982: A Biographical Memoir* (Washington, DC: National Academy of Sciences, 1994), p. 216.
2. Mark Harrison, "Soviet Economic Growth Since 1928: The Alternative Statistics of G. I. Khanin," *Europe–Asia Studies* 45, no. 1 (1993): 141–67.
3. Wassily Leontief thought in a similar way, writing, "We are accustomed to rewarding people for work based on market mechanisms, but we can no longer rely on the market mechanism to function so conveniently." Quoted in Timothy Taylor, "Automation and Job Loss: Leontief in 1982," 22 August 2016, available at http://conversableeconomist.blogspot.com/2016/08/automation-and-job-loss-leontief-in-1982.html (accessed February 2019).
4. Walter Scheidel, *The Great Leveler: Violence and the History of Inequality from the Stone Age to the Twenty-First Century* (Oxford: Princeton University Press, 2017).
5. Phillippe Van Parijs and Yannick Vanderborght, *Basic Income: A Radical Proposal for a Free Society and a Sane Economy* (London: Harvard University Press, 2017), p. 52.
6. Anthony B. Atkinson, *Inequality: What Can Be Done?* (London: Harvard University Press, 2015), p. 264; "The Welfare State Needs Updating," *Economist*, 12 July 2018.

7. See Nicholas Timmins, "Commission on Social Justice," *Independent*, 25 October 1994; and "The Welfare State Needs Updating." The Beveridge Report is also partly responsible for one of the great upsets in modern political history. Winston Churchill, the prime minister who carried Britain to victory in the Second World War, fell in the election he called only two months after the war ended. His opponent, Clement Atlee, was thought to be a better custodian of the ideas in the report.
8. Chapter 2 in Van Parijs and Vanderborght, *Basic Income*, sets out these proposals in great detail.
9. Critics, for their part, have referred to safety nets as "hammocks." See "The Welfare State Needs Updating."
10. William Beveridge, *Social Insurance and Allied Services* (London: Her Majesty's Stationery Office, 1942), p. 6.
11. Peter Diamond and Emmanuel Saez, "The Case for a Progressive Tax: From Basic Research to Policy Recommendations," *Journal of Economic Perspectives* 25, no. 4 (2011): 165–90.
12. See, for instance, Thomas Piketty and Emmanuel Saez, "A Theory of Optimal Capital Taxation," NBER Working Paper No. 17989 (2012).
13. Ibid., p. 1.
14. Kevin J. Delaney, "The Robot That Takes Your Job Should Pay Taxes, Says Bill Gates," *Quartz*, 17 February 2017.
15. Blaine Harden, "Recession Technology Threaten Workers," *Washington Post*, 26 December 1982.
16. International Association of Machinists, "Workers' Technology Bill of Rights," *Democracy* 3, no. 1 (1983). That same Machinists Union remains in action today, striking a deal with Uber that provides its New York drivers with a spread of benefits, for example. Leslie Hook, "Uber Strikes Deal with Machinists Union for New York Drivers," *Financial Times*, 10 May 2016.
17. Lawrence Summers, "Robots Are Wealth Creators and Taxing Them Is Illogical," *Financial Times*, 5 March 2017.
18. Both where the traditional capital is saved (the "stock") and the income that streams to those who own it (the "flow"). Piketty similarly called for a global tax on capital; see Piketty, *Capital in the Twenty-First Century*.
19. Figure 10.1 is for those with a GDP greater than $300 billion in 2007 (from Annette Alstadsæter, Niels Johannesen, and Gabriel Zucman, "Who Owns the Wealth in Tax Havens? Macro Evidence and Implications for Global Inequality," *Journal of Public Economics* 162 (2018): 89–100. The figure was produced by Gabriel Zucman, available at https://gabriel-zucman.eu/offshore/ (accessed September 2018).
20. Alstadsæter, Johannesen, and Zucman, "Who Owns the Wealth in Tax Havens?," p. 100.
21. This reasoning is from James Mirrlees and Stuart Adam, *Dimensions of Tax Design: The Mirrlees Review* (Oxford: Oxford University Press, 2010), p. 757.
22. Alstadsæter, Johannesen, and Zucman, "Who Owns the Wealth in Tax Havens?," figure 5. Figure is adapted from one produced by Gabriel Zucman, available at https://gabriel-zucman.eu/offshore/ (accessed September 2018).
23. "Taxing Inheritances Is Falling Out of Favour," *Economist*, 23 November 2017.
24. Ibid.; and Caroline Freund and Sarah Oliver, "The Origins of the Superrich: The

Billionaire Characteristics Database," *Peterson Institute for International Economics* 16, no. 1 (2016).

25. David Autor, David Dorn, Lawrence Katz, et al., "The Fall of the Labor Share and the Rise of Superstar Firms," NBER Working Paper No. 23396 (2017); and Simcha Barkai, "Declining Labor and Capital Shares," Working Paper, University of Chicago (2016).
26. Lynnley Browning and David Kocieniewski, "Pinning Down Apple's Alleged 0.005% Tax Rate Is Nearly Impossible," *Bloomberg*, 1 September 2016, quoted in Daron Acemoglu and Simon Johnson, "It's Time to Found a New Republic," *Foreign Policy*, 15 August 2017. See https://ec.europa.eu/ireland/tags/taxation_en on Apple as well. The basic tax rate in the Republic of Ireland is 20 percent.
27. This share has gone from 2 to 17 percent; see figure 3 in Gabriel Zucman, "Taxing Across Borders: Tracking Personal Wealth and Corporate Profits," *Journal of Economic Perspectives* 28, no. 4 (2014): 121–48.
28. "Apple Pays Disputed Irish Tax Bill," BBC News, 18 September 2018.
29. The traditional view of lawyers is set out in a well-known case between the Duke of Westminster and the Inland Revenue Commissioners, decided in 1936. One of the then Law Lords, Lord Tomlin, gave judgment in favour of a scheme adopted by the taxpayer. He said this: "Every man is entitled if he can to arrange his affairs so that the tax attaching under the appropriate Acts is less than it otherwise would be. If he succeeds in ordering them so as to secure that result, then, however unappreciative the Commissioners of Inland Revenue or his fellow taxpayers may be of his ingenuity, he cannot be compelled to pay an increased tax." *IRC v. Duke of Westminster* [1936] AC1 (HL).
30. Following Holmes, the US Internal Revenue Service has inscribed "Taxes are what we pay for a civilized society" above the entrance to their headquarters in Washington, DC; see https://quoteinvestigator.com/2012/04/13/taxes-civilize/.
31. This is figure 5 in Zucman, "Taxing Across Borders." The figure is adapted from one available at https://gabriel-zucman.eu/ (accessed September 2018). This figure reports decennial averages (e.g., 1990–99 is the average of 1990, 1991 . . . and 1999).
32. TRAC, "Millionaires and Corporate Giants Escaped IRS Audits in FY 2018," published online at https://trac.syr.edu/tracirs/latest/549/ (accessed May 8, 2019). Thanks to Adam Tooze for bringing this to my attention.
33. Thomas Paine, *Agrarian Justice* (Digital Edition, 1999), available at http://piketty.pse.ens.fr/%EF%AC%81les/Paine1795.pdf.
34. Van Parijs and Vanderborght, *Basic Income*.
35. See Victor Oliveira, "The Food Assistance Landscape," Economic Research Service at the United States Department of Agriculture, Economic Information Bulletin Number 169 (March 2017). It is worth $125.51 per month per person, so about $1,506 per annum.
36. For health care, see https://www.nuffieldtrust.org.uk/chart/health-spending-per-person-in-england-dh-and-nhs-england (accessed 24 April 2018). For education, see Chris Belfield, Claire Crawford, and Luke Sibieta, "Long-Run Comparisons for Spending per Pupil Across Different Stages of Education," Institute for Fiscal Studies, 27 February 2017. In 2016, £2,215 of health care was spent per person per annum, and £4,900 primary education and £6,300 secondary education was spent per person per annum.
37. Galbraith, *The Affluent Society*, p. 239.
38. Friedrich Hayek, quoted in Van Parijs and Vanderborght, *Basic Income*, p. 86.

39. Lowrey, *Give People Money*; Chris Hughes, *Fair Shot: Rethinking Inequality and How We Earn* (London: Bloomsbury, 2018).
40. The quotation is from Paine, *Agrarian Justice*. The calculation is from Atkinson, *Inequality: What Can Be Done?* p. 169.
41. The idea of an "admissions policy" is an important part of the argument in Avent, *The Wealth of Humans*. See, for instance, the introduction, where Avent eloquently explains how "fights about who belongs *within* particular societies . . . will also intensify."
42. See https://www.bia.gov/frequently-asked-questions (accessed 23 April 2018).
43. See "American Indian and Alaska Native Heritage Month: November 2017" at https://www.census.gov/newsroom/facts-for-features/2017/aian-month.html; and "Suicide: 2016 Facts & Figures," American Foundation for Suicide Prevention, at https://afsp.org/ (accessed September 2018).
44. See Gross Gaming Revenue Reports, National Indian Gaming Commission, https://www.nigc.gov/commission/gaming-revenue-reports; "Of Slots and Sloth," *Economist*, 15 January 2015; and Cecily Hilleary, "Native American Tribal Disenrollment Reaching Epidemic Levels," *VOA*, 3 March 2017. I first came across this case in extracts from a conversation between David Autor and Tim O'Reilly, "Work Is More Than a Source of Income," *Medium*, 28 September 2015.
45. "The Welfare State Needs Updating."
46. Phillippe Van Parijs, "Basic Income: A Simple and Powerful Idea for the Twenty-First Century" in Bruce Ackerman, Anne Alstott, and Phillipe Van Parijs, eds., *Redesigning Distribution: Basic Income and Stakeholder Grants as Cornerstones for an Egalitarian Capitalism* (New York: Verso, 2005), p. 14.
47. Winning the lottery, for instance, appears to make people work less. That suggests that other types of unearned income, like a UBI, might have similar effects. But in 2010, when the Iranian government made direct cash payments worth about 29 percent of median income to more than 70 million people, it did not appear to have that negative effect on work. (President Ahmadinejad felt forced to do this when he removed subsidies on energy and bread prices.) Large inheritances also appear to encourage people not to work—those who receive $150,000 are four times more likely to leave the labor market than those who receive less than $25,000. Again, that suggests that unearned income like a UBI might make work a less attractive proposition. Yet a different study looking at the $2,000 payments made annually to all Alaskan residents by the Alaskan Permanent Fund (on which more later in this chapter) found that it had no effect on employment. See Guido Imbens, Donald Rubin, and Bruce Sacerdote, "Estimating the Effect of Unearned Income on Labor Earnings, Savings, and Consumption: Evidence from a Survey of Lottery Players," *American Economic Review* 91, no. 4 (2001): 778–94; Djaved Salehi-Isfahani and Mohammad Mostafavi-Dehzooei, "Cash Transfers and Labor Supply: Evidence from a Large-Scale Program in Iran," *Journal of Development Economics* 135 (2018): 349–67; Douglas Holtz-Eakin, David Joulfaian, and Harvey Rosen, "The Carnegie Conjecture: Some Empirical Evidence," *Quarterly Journal of Economics* 108, no. 2 (1993): 413–35; Damon Jones and Ioana Marinescu, "The Labor Market Impact of Universal and Permanent Cash Transfers: Evidence from the Alaska Permanent Fund," NBER Working Paper No. 24312 (February 2018).
48. Jon Elster, "Comment on Van der Veen and Van Parijs," *Theory and Society* 15, no. 5 (1986): 709–21.

49. Alberto Alesina, Reza Baqir, and William Easterly, "Public Goods and Ethnic Divisions," *Quarterly Journal of Economics* 114, no. 4 (1999): 1243–84.
50. Alberto Alesina, Edward Glaeser, and Bruce Sacerdote, "Why Doesn't the United States Have a European-Style Welfare State?," *Brookings Papers on Economic Activity* 2 (2001).
51. John Lloyd, "Study Paints Bleak Picture of Ethnic Diversity," *Financial Times*, 8 October 2006.
52. Tom Bartlett, "Harvard Sociologist Says His Research Was 'Twisted,'" *Chronicle of Higher Education*, 15 August 2012.
53. Michael Sandel, "Themes of 2016: Progressive Parties Have to Address the People's Anger," *Guardian*, 1 January 2017. He also asks what the "moral significance" of "national borders" in particular might be.
54. I am grateful to several Balliol undergraduate students of philosophy, politics, and economics for bringing the field of contributive justice to my attention in their writing.
55. See both Van Parijs and Vanderborght, *Basic Income*, p. 29; and Atkinson, *Inequality: What Can Be Done?* This proposal travels under a variety of names—a "national youth endowment" and a "minimum inheritance," a "universal personal capital account" and a "stakeholder grant," a "child trust fund" and a "capital endowment."
56. See Will Kymlicka, *Contemporary Political Philosophy: An Introduction* (New York: Oxford University Press, 2002), p. 170, where he writes that "if all we do is redistribute income from those who own productive assets to those who do not, then we will still have classes, exploitation, and hence the kind of contradictory interests that make justice necessary in the first place. We should instead be concerned with transferring ownership of the means of production themselves. When this is accomplished, questions of fair distribution become obsolete."
57. Joshua Brustein, "Juno Sold Itself as the Anti-Uber. That Didn't Last Long," *Bloomberg*, 28 April 2017.
58. Susskind and Susskind, *Future of the Professions*, p. 34.
59. Jesse Bricker, Lisa J. Dettling, Alice Henriques, et al., "Changes in U.S. Family Finances from 2013 to 2016: Evidence from the Survey of Consumer Finances," *Federal Reserve Bulletin* 103, no. 3 (2017).
60. See https://www.nbim.no/en/the-fund/about-the-fund/ (accessed October 2018).
61. This account is found in Chris Hughes, *Fair Shot: Rethinking Inequality and How We Earn* (London: Bloomsbury, 2018), p. 137.
62. This is a reproduction of figure 3.1.3 in Alvaredo, Chancel, Piketty, et al., *World Inequality Report.*
63. "Why Trade Unions Are Declining," *Economist*, 29 September 2015.
64. John Kenneth Galbraith, *American Capitalism: The Concept of Countervailing Power* (Eastford, CT: Martino Fine Books, 2012).
65. Satya Nadella, "The Partnership of the Future," *Slate*, 28 June 2016.
66. Kevin Roose, "The Hidden Automation Agenda of the Davos Elite," *New York Times*, 25 January 2019.
67. Ryan Abbott and Bret Bogenschneider, "Should Robots Pay Taxes? Tax Policy in the Age of Automation," *Harvard Law & Policy Review* 12 (2018).
68. "New Poll Reveals 8 in 10 Londoners Believe Capital's Nurses Are Underpaid," Royal College of Nursing, 6 September 2017, https://www.rcn.org.uk/; "The 50th Annual

PDK Poll of the Public's Attitudes Toward the Public Schools, Teaching: Respect but Dwindling Appeal," PDK Poll, http://pdkpoll.org/ (accessed September 2018).

69. Guy Chazan, "German Union Wins Right to 28-Hour Working Week and 4.3% Pay Rise," *Financial Times*, 6 February 2018.
70. See https://www.tuc.org.uk/about-the-tuc; Rebecca Wearn, "Unions Call for Four-day Working Week," BBC News, 10 September 2018.
71. "Technology May Help to Revive Organised Labour," *Economist*, 15 November 2018.
72. Frances O'Grady, "Building a Movement Fit for the Next 150 Years," 10 September 2018, available at https://www.tuc.org.uk/blogs/ (accessed September 2018).
73. Alexandra Topping, "Frances O'Grady on the TUC at 150: 'Unions Have to Change or Die'," *Guardian*, 4 June 2018.

11. BIG TECH

1. "62.6 percent" from https://www.comscore.com/Insights/Rankings?country=US (accessed 1 May 2019); "88 percent" from Jonathan Taplin, "Is It Time to Break Up Google?," *New York Times*, 22 April 2017.
2. Facebook has 2.38 billion monthly active users as of April 2019, with a global population of ~7.7 billion. See https://newsroom.fb.com/company-info/ (accessed 1 May 2019) and https://en.wikipedia.org/wiki/World_population (accessed 1 May 2019). For the "77 percent" and "74 percent," see Taplin, "Is It Time to Break Up Google?" For the "43 percent," see "Amazon Accounts for 43% of US Online Retail Sales," *Business Insider Intelligence*, 3 February 2017.
3. Greg Ip, "The Antitrust Case Against Facebook, Google and Amazon," *Wall Street Journal*, 16 January 2018.
4. PwC, "Global Top 100 Companies by Market Capitalisation" (2018).
5. Marc Andreessen, "Why Software Is Eating the World," *Wall Street Journal*, 20 August 2011.
6. Connie Chan, "When One App Rules Them All: The Case of WeChat and Mobile in China," Andreessen Horowitz, https://a16z.com/2015/08/06/wechat-china-mobile-first/, quoted in J. Susskind, *Future Politics*, p. 331.
7. Dan Frommer, "Microsoft Is Smart to Prepare for Its New Role as Underdog," *Quartz*, 17 July 2014.
8. James Klisner, "IBM: Creating Shareholder Value with AI? Not So Elementary, My Dear Watson," *Jefferies Franchise Note*, 12 July 2017. See https://javatar.bluematrix.com/pdf/fO5xcWjc.
9. Avery Hartmans, "These 18 Incredible Products Didn't Exist 10 Years Ago," *Business Insider UK*, 16 July 2017.
10. Andre Esteva, Brett Kuprel, Roberto A. Novoa, et al., "Dermatologist-Level Classification of Skin Cancer with Deep Neural Networks," *Nature* 542 (2017): 115–18.
11. See Jeff Reinke, "From Old Steel Mill to Autonomous Vehicle Test Track," *Thomas*, 19 October 2017; Michael J. Coren, "Tesla Has 780 Million Miles of Driving Data, and Adds Another Million Every 10 Hours," *Quartz*, 28 May 2016; and Alexis C. Madrigal, "Inside Waymo's Secret World for Training Self-Driving Cars," *Atlantic*, 23 August 2017.
12. David McCandless, "Codebases: Millions of Lines of Code," 24 September 2015,

https://informationisbeautiful.net/visualizations/million-lines-of-code/ (accessed 25 April 2018).

13. Michael J. Coren, "San Francisco Is Actually One of the Worst-Paying Places in the US for Software Engineers," *Quartz*, 9 February 2017.
14. Udi Manber and Peter Norvig, "The Power of the Apollo Missions in a Single Google Search," *Google Inside Search*, 28 August 2012, https://search.googleblog.com/2012/08/the-power-of-apollo-missions-in-single.html (accessed 25 April 2018).
15. Satinder Singh, Andy Okun, and Andrew Jackson, "Artificial Intelligence: Learning to Play Go from Scratch," *Nature* 550 (2017): 336–37.
16. Silver, Schrittwieser, Simonyan, et al., "Mastering the Game of Go Without Human Knowledge."
17. J. Susskind, *Future Politics*, p. 318.
18. Massimo Motta, *Competition Policy* (Cambridge: Cambridge University Press, 2007), p. 39.
19. Peter Thiel, "Competition Is for Losers," *Wall Street Journal*, 12 September 2014.
20. For Microsoft, see Joel Brinkley, "U.S. vs. Microsoft: The Overview; U.S. Judge Says Microsoft Violated Antitrust Laws with Predatory Behavior," *New York Times*, 4 April 2000; for Facebook, see Guy Chazan, "German Antitrust Watchdog Warns Facebook over Data Collection," *Financial Times*, 19 December 2017; for Google, see Rochelle Toplensky, "Google Appeals €2.4bn EU Antitrust Fine," *Financial Times*, 11 September 2017; for Apple, see Adam Liptak and Vindu Goel, "Supreme Court Declines to Hear Apple's Appeal in E-Book Pricing Case," *New York Times*, 7 March 2011; for Amazon, see Simon van Dorpe, "The Case Against Amazon," *Politico.eu*, 4 March 2019.
21. This example is set out in Peter Thiel and Blake Masters, *Zero to One* (New York: Crown Business, 2014).
22. Michael Cox, "Schumpeter in His Own Words," *Federal Reserve Bank of Dallas: Economic Insights* 6, no. 3 (2001): 5.
23. Joseph A. Schumpeter, *Capitalism, Socialism, and Democracy* (London: Routledge, 2005).
24. Mark J. Perry, "Fortune 500 Firms 1955 v. 2017: Only 60 Remain, Thanks to the Creative Destruction That Fuels Economic Prosperity," *AEIdeas*, 13 October 2017.
25. See Ariel Ezrachi and Maurice Stucke, *Virtual Competition: The Promise and Perils of the Algorithm-Driven Economy* (Cambridge, MA: Harvard University Press, 2016), chaps. 5–8.
26. Emilio Calvano, Giacomo Calzolari, Vincenzo Denicolò, and Sergio Pastorello, "Artificial Intelligence, Algorithmic Pricing, and Collusion," Vox CEPR Policy Portal, 3 February 2019, available at https://voxeu.org/ (accessed February 2019).
27. Benedict Evans, "Why Amazon Has No Profits (And Why It Works)," 5 September 2014, https://www.ben-evans.com/benedictevans/2014/9/4/why-amazon-has-no-profits-and-why-it-works (accessed 25 April 2018).
28. Data extracted from Rani Molla and Jason Del Ray (2017), "Amazon's Epic 20-Year Run as a Public Company Explained in Five Charts," recode.net, May 15, updated from Q4 2017 using Amazon data at https://ir.aboutamazon.com/quarterly-results.
29. Timothy Lee, "Uber Lost $2.8 Billion in 2016. Will It Ever Become Profitable?," *Vox*, 15 April 2017.
30. Lina M. Khan, "Amazon's Antitrust Paradox," *Yale Law Journal* 126, no. 3 (2017): 564–907.

31. See the history of ExxonMobil, for instance, at http://corporate.exxonmobil.com/en/company/about-us/history/overview (accessed 25 April 2018).
32. "May 15, 1911: Supreme Court Orders Standard Oil to Be Broken Up," *New York Times*, 15 May 2012.
33. "The World's Most Valuable Resource Is No Longer Oil, but Data," *Economist*, 6 May 2017.
34. See *Standard Oil Co. of New Jersey v. United States*, 221 U.S. 1 (1911), available at https://supreme.justia.com/cases/federal/us/221/1/case.html (accessed 25 April 2018).
35. "Google Searches Expose Racial Bias, Says Study of Names," BBC News, 4 February 2013, quoted in J. Susskind, *Future Politics*, p. 288.
36. James Vincent, "Google 'Fixed' Its Racist Algorithm by Removing Gorillas from Its Image-Labeling Tech," *Verge*, 12 January 2018, quoted in J. Susskind, *Future Politics*, p. 282.
37. Ernesto, "Google Removed 2.5 Billion 'Pirate' Search Results," *TorrentFreak*, 6 July 2017.
38. Craig Silverman, "Recommendations Push Users to the Fringe," *BuzzFeed*, 12 April 2018; Caroline O'Donovan, "YouTube Just Demonetized Anti-Vax Channels," *BuzzFeed*, 22 February 2019; Eli Rosenberg, "A Right-Wing YouTuber Hurled Racist, Homophobic Taunts at a Gay Reporter. The Company Did Nothing," *Washington Post*, 5 June 2019; Max Fisher and Amanda Taub, "On YouTube's Digital Playground, an Open Gate for Pedophiles," *New York Times*, 3 June 2019.
39. Robert Booth, "Facebook Reveals News Feed Experiment to Control Emotions," *Guardian*, 30 June 2014.
40. J. Susskind, *Future Politics*, p. 132.
41. Karen Hao, "Facebook's Ad-Serving Algorithm Discriminates by Gender and Race," *MIT Technology Review*, 5 April 2019.
42. Scott Shane, "These Are the Ads Russia Bought on Facebook in 2016," *New York Times*, 1 November 2017; Eric Tucker and Mary Clare Jalonick, "Lawmakers Release Troves of Facebook Ads Showing Russia's Cyber Intrusion," *Chicago Tribune*, 1 November 2017.
43. Karsten Müller and Carlo Schwarz, "Fanning the Flames of Hate: Social Media and Hate Crime," Warwick University Working Paper Series No. 373 (May 2018), reported in Amanda Taub and Max Fisher, "Facebook Fueled Anti-refugee Attacks in Germany, New Research Suggests," *New York Times*, 21 August 2018.
44. Brad Stone, "Amazon Erases Orwell Books from Kindle," *New York Times*, 17 July 2009.
45. James Felton, "Amazon AI Designed to Create Phone Cases Goes Hilariously Wrong," *IFLScience!*, 10 July 2017.
46. Nicole Chavez, "Arkansas Judge Drops Murder Charge in Amazon Echo Case," *CNN*, 2 December 2017.
47. J. Susskind, *Future Politics*, p. 236; the "Inconvenient Facts" app, at https://apps.apple.com/us/app/inconvenient-facts/id1449892823?ls=1 (accessed June 2019); Josh Begley, "After 12 Rejections, Apple Accepts App That Tracks U.S. Drone Strikes," *Intercept*, 28 March 2017; Ben Hubbard, "Apple and Google Urged to Dump Saudi App That Lets Men Track Women," *New York Times*, 13 February 2019.
48. Arash Khamooshi, "Breaking Down Apple's iPhone Fight with the U.S. Government," *New York Times*, 21 March 2016. Also see J. Susskind, *Future Politics*, p. 155.

49. James Vincent, "Twitter Taught Microsoft's AI Chatbot to be a Racist Asshole in Less Than a Day," *Verge*, 24 March 2016. Also see J. Susskind, *Future Politics*, p. 37.
50. Alex Kantrowitz, "Microsoft's Chatbot Zo Calls the Qur'an Violent and Has Theories About Bin Laden," *BuzzFeed News*, 3 July 2017.
51. J. Susskind, *Future Politics*, p. 73.
52. Ibid., p. 3.
53. On the privatization of political life, see Jamie Susskind, "Future Politics: Living Together in a World Transformed by Tech," Google Talks, 18 October 2018, https://www.youtube.com/watch?v=PcPJjOJO1vo (accessed October 2018).
54. Tony Romm, Drew Harwell, and Craig Timberg, "Google CEO Sundar Pichai: Fears About Artificial Intelligence Are 'Very Legitimate,' He Says in Post Interview," *Washington Post*, 12 December 2018.
55. See, for instance, Nick Srnicek, "We Need to Nationalise Google, Facebook, and Amazon. Here's Why," *Guardian*, 30 August 2017; Nick Srnicek, "The Only Way to Rein in Big Tech Is to Treat Them as a Public Service," *Guardian*, 23 April 2019.
56. Simon Mistreanu, "Life Inside China's Social Credit Laboratory," *Foreign Policy*, 3 April 2018.
57. Christopher Hope, "One Official Disciplined over Data Loss Every Day," *Telegraph*, 3 November 2008.
58. On "consumers" versus "citizens," see Jamie Susskind, "Future Politics: Living Together in a World Transformed by Tech," Harvard University CLP Speaker Series, 11 December 2018.

12. MEANING AND PURPOSE

1. This was a point made by Michael Sandel in "In Conversation with Michael Sandel: Capitalism, Democracy, and the Public Good," LSE Public Lecture chaired by Tim Besley, 2 March 2017, http://www.lse.ac.uk/ (accessed April 24, 2018).
2. Quoted in David Spencer, *The Political Economy of Work*, digital ed. (New York: Routledge, 2010), p. 19.
3. Both quoted ibid., p. 79.
4. See https://www.amazon.com/Love-work-love-thats-all/dp/B01M0EY8ZD (accessed 24 April 2018).
5. See https://www.freud.org.uk/about/faq/ (accessed 19 October 2017); Sigmund Freud, *Civilization and Its Discontents* (New York: W. W. Norton, 2010), pp. 79–80.
6. Max Weber, *The Protestant Ethic and the Spirit of Capitalism* (Oxford: Oxford University Press, 2011), p. 129.
7. Ibid., p. 170.
8. Ibid., pp. 99–100.
9. Marie Jahoda, Paul Lazarsfeld, and Hans Zeisel, *Marienthal: The Sociography of an Unemployed Community* (Piscataway, NJ: Transaction Publishers, 2009), p. vii. The following account of the Marienthal Study is all drawn from this book.
10. See, for instance, Marie Jahoda, *Employment and Unemployment: A Social-Psychological Analysis* (Cambridge: Cambridge University Press, 1982); on suicides, see "Why Suicide Is Falling Around the World, and How to Bring It Down More," *Economist*, 24 November 2018.

11. Michael Sandel, "Themes of 2016: Progressive Parties Have to Address the People's Anger," *Guardian*, 1 January 2017.
12. Sandel, "In Conversation with Michael Sandel."
13. Norman Longmate, *The Workhouse: A Social History* (London: Pimlico, 2003), p. 14.
14. Chris Weller, "EBay's Founder Just Invested $500,000 in an Experiment Giving Away Free Money," *Business Insider UK*, 8 February 2017.
15. Clark, *Farewell to Alms*, p. 65.
16. Ibid., p. 66.
17. Ibid., pp. 64–65.
18. James Suzman, *Affluence Without Abundance: The Disappearing World of the Bushmen* (London: Bloomsbury, 2017), p. 256.
19. Hannah Arendt, *The Human Condition* (London: University of Chicago Press, 1998), p. 82.
20. Aristotle, *Politics*, book III, available at http://www.perseus.tufts.edu/.
21. James Renshaw, *In Search of the Greeks* (London: Bloomsbury, 2015), p. 376.
22. Aristotle, quoted in J. Susskind, *Future Politics*, p. 301. This particular translation is from Kory Schaff, *Philosophy and the Problems of Work: A Reader* (Oxford: Rowman & Littlefield, 2001).
23. Maurice Balme, "Attitudes to Work and Leisure in Ancient Greece," *Greece & Rome* 31, no. 2 (1984): 140–52.
24. Jacob Snyder, "Leisure in Aristotle's Political Thought," *Polis: The Journal for Ancient Greek Political Thought* 35, no. 2 (2018).
25. Cited in Balme, "Attitudes to Work," but originally from Hesiod, *Theogony, Works and Days, Testimonia*, "Theogony," lines 535–57.
26. Hesiod, "Works and Days," lines 42–53, in *Theogony, Works and Days, Testimonia*, Most, ed. and trans.
27. Cited in Balme, "Attitudes to Work"; Genesis 3:19, https://www.biblegateway.com/.
28. See, for instance, "Economic and Philosophical Manuscripts," in Karl Marx, *Selected Writings*, ed. Lawrence Simon (Indianapolis: Hackett, 1994).
29. Quoted in Susskind and Susskind, *Future of the Professions*, p. 256.
30. Fourier is quoted in David Frayne, *The Refusal of Work: The Theory and Practice of Resistance to Work* (London: Zed Books, 2015), p. 30.
31. Susskind and Susskind, *Future of the Professions*, p. 255.
32. Gallup, "State of the American Workplace" (2017); Pew Research Center, "How Americans View Their Jobs," 6 October 2016, http://www.pewsocialtrends.org/2016/10/06/3-how-americans-view-their-jobs/ (accessed 24 April 2018).
33. Will Dahlgreen, "37% of British Workers Think Their Jobs Are Meaningless," YouGov UK, 12 August 2015.
34. David Graeber, "On the Phenomenon of Bullshit Jobs: A Work Rant," *STRIKE!* magazine, August 2013.
35. Pierre-Michel Menger calls this "the French Paradox." He set it out in a presentation titled "What Is Work Worth (in France)?," prepared for the "Work in the Future" symposium, 6 February 2018, organized by Robert Skidelsky.
36. Keynes, *Essays in Persuasion*, p. 368.
37. Leontief, "National Perspective," p. 7.

38. Bertrand Russell, *In Praise of Idleness and Other Essays* (New York: Routledge, 2004), pp. 3 and 13.
39. Thorstein Veblen, *The Theory of the Leisure Class* (New York: Dover Thrift Editions, 1994).
40. G. A. Cohen, *If You're an Egalitarian, How Come You're So Rich?* (London: Harvard University Press, 2001).
41. See http://www.english-heritage.org.uk/learn/story-of-england/victorian/religion/ (accessed 24 April 2018).
42. Kory Schaff, *Philosophy and the Problems of Work: A Reader* (Oxford: Rowman & Littlefield, 2001), p. 3; Keynes, *Essays in Persuasion*, p. 368.
43. Jahoda, Lazarsfeld, and Zeisel, *Marienthal*, p. 66.
44. Eleanor Dickey, "Education, Research, and Government in the Ancient World," lecture at Gresham College, Barnard's Inn Hall, London, 15 May 2014.
45. Michael Barber, "Rab Butler's 1944 Act Brings Free Secondary Educational for All," *BBC News*, 17 January 1944.
46. See Jonathan Birdwell, Ralph Scott, and Louis Reynolds, *Character Nation* (London: Demos, 2015), p. 9.
47. James Arthur, Kristján Kristjánsson, David Walker, et al., "Character Education in UK Schools Research Report," The Jubilee Centre for Character and Virtues at the University of Birmingham (2015), as described ibid., p. 10.
48. Men, 6.1 × 7 = 42.7 hours a week; women, 38.5 hours a week. Office for National Statistics, "Leisure Time in the UK: 2015," 24 October 2017, https://www.ons.gov.uk/releases/leisuretimeintheuk2015 (accessed 24 April 2017). We should take TV's supposed primacy with a pinch of salt, though: the ONS classification does not appear to properly capture time spent online.
49. See http://www.bbc.co.uk/corporate2/insidethebbc/whoweare/mission_and_values (accessed 8 May 2018). In a recent dispute with the BBC, the chairman of the UK Conservative Party threatened to cut off its funding in just this fashion. See Tim Ross, "BBC Could Lose Right to Licence Fee over 'Culture of Waste and Secrecy,' Minister Warns," *Telegraph*, 26 October 2013.
50. HM Government, "Sporting Future: A New Strategy for an Active Nation," December 2015.
51. Sarah O'Connor, "Retirees Are Not the Only Ones Who Need a Break," *Financial Times*, 7 August 2018.
52. The volunteering statistics are from Andy Haldane, "In Giving, How Much Do We Receive? The Social Value of Volunteering," lecture to the Society of Business Economists, London, 9 September 2014. That year, there were 30.8 million people in paid work in the UK. See Office for National Statistics, "Statistical Bulletin: UK Labour Market, December 2014," 17 December 2014.
53. Haldane, "In Giving, How Much Do We Receive?"
54. Sophie Gilbert, "The Real Cost of Abolishing the National Endowment for the Arts," *Atlantic*, 16 March 2017.
55. For the UK, Daniel Wainwright, Paul Bradshaw, Pete Sherlock, and Anita Geada, "Libraries Lose a Quarter of Staff as Hundreds Close," BBC News, 29 March 2016—4,290 council-run libraries in 2010, 3,765 in 2016. Interestingly, this story is not universal—in China, over a similar period, the number of public libraries increased

by 8.4 percent. See Will Dunn, "The Loss of Britain's Libraries Could Be a Huge Blow to the Economy," *New Statesman*, 18 December 2017.

56. Benedicte Page, "Philip Pullman's Call to Defend Libraries Resounds Around the Web," *Guardian*, 27 January 2011.
57. Orrin E. Dunlap Jr., "Telecasts to Homes Begin on April 30—World's Fair Will Be the Stage," *New York Times*, 19 March 1939.
58. "Ulysses" in Alfred Tennyson, *Selected Poems* (London: Penguin Books, 2007).
59. Dylan Matthews, "4 Big Questions About Job Guarantees," *Vox*, 27 April 2018; Sean McElwee, Colin McAuliffe, and Jon Green, "Why Democrats Should Embrace a Federal Jobs Guarantee," *Nation*, 20 March 2018.
60. See, for instance, Arendt, *Human Condition*, on "labor," "work," and "action." Many socialists, meanwhile, hope that in the future the distinction might melt away altogether, as work becomes leisure, leisure becomes work, both become, in Marx's words, "not only a means of life but life's prime want." See "Critique of the Gotha Program" in Marx, *Selected Writings*, p. 321.
61. "We are not the only people who consider the man who takes no part in politics not as one who minds his own business but as useless," from Pericles's Funeral Oration, quoted in Balme, "Attitudes to Work."
62. International Labour Organization, *Care Work and Care Jobs for the Future of Decent Work* (Geneva: International Labour Office, 2018), p. xxvii.
63. Lowrey, *Give People Money*, p. 151.
64. "Unpaid Care," Parliamentary Office of Science and Technology, Houses of Parliament, No. 582 (July 2018).
65. See Chris Rhodes, "Manufacturing: Statistics and Policy," House of Commons Library Brief Paper No. 01942 (November 2018); Chris Payne and Gueorguie Vassilev, "House Satellite Account, UK: 2015 and 2016," Office for National Statistics (October 2018). GVA of manufacturing sector in 2016 was £176bn; of "household housing services," "nutrition," "laundry," and "child care," £797.65bn.
66. Joi Ito and Scott Dadich, "Barack Obama, Neural Nets, Self-Driving Cars, and the Future of the World," *Wired*, 12 October 2016.
67. Alex Moss, "Kellingley Mining Machines Buried in Last Deep Pit," *BBC News*, 18 December 2015.
68. See, for instance, David Goodhart and Eric Kaufmann, "Why Culture Trumps Skills: Public Opinion on Immigration," Policy Exchange, 28 January 2018.
69. John Stuart Mill, *Principles of Political Economy with Chapters on Socialism* (Oxford: Oxford University Press, 2008), p. 124.
70. Isaiah Berlin, *Two Concepts of Liberty* (Oxford: Clarendon Press, 1958), p. 3.

EPILOGUE

1. Stefan Zweig, *The World of Yesterday* (London: Pushkin Press, 2014), p. 23.

BIBLIOGRAPHY

The following is a list of all books, scholarly articles, and lectures referenced in the text and contained in the notes, as well as some general articles of significance. For websites and online data tables, please refer to the notes. Here and in the notes, the dates of last access to the websites are given only when the references relate to data and facts that might be subject to change over time.

Abbott, Ryan, and Bret Bogenschneider. "Should Robots Pay Taxes? Tax Policy in the Age of Automation." *Harvard Law & Policy Review* 12 (2018).

Acemoglu, Daron. "Technical Change, Inequality, and the Labor Market." *Journal of Economic Literature* 40, no. 1 (2002): 7–72.

Acemoglu, Daron, and David Autor. "Skills, Tasks and Technologies: Implications for Employment and Earnings." In David Card and Orley Ashenfelter, eds., *Handbook of Labor Economics*, vol. 4, pt. B (North-Holland: Elsevier, 2011), pp. 1043–171.

Acemoglu, Daron, and Pascual Restrepo. "Artificial Intelligence, Automation and Work." In Ajay Agrawal, Joshua Gans, and Avi Goldfarb, eds., *Economics of Artificial Intelligence* (Chicago: Chicago University Press, 2018).

———. "Demographics and Automation." NBER Working Paper No. 24421 (2018).

———. "The Race Between Machine and Man: Implications of Technology for Growth, Factor Shares, and Employment." *American Economic Review* 108, no. 6 (2018): 1488–542.

———. "Robots and Jobs: Evidence from US Labor Markets." NBER Working Paper No. 23285 (2017).

———. "The Wrong Kind of AI? Artificial Intelligence and the Future of Labor Demand." MIT Working Paper (2019).

Acemoglu, Daron, and James Robinson. *Why Nations Fail*. London: Profile Books, 2012.

Adams, James Truslow. *The Epic of America*. New York: Little, Brown, 1931.

Albers, Anni. *On Weaving*. Princeton, NJ: Princeton University Press, 2017.

Alesina, Alberto, Reza Baqir, and William Easterly. "Public Goods and Ethnic Divisions." *Quarterly Journal of Economics* 114, no. 4 (1999): 1243–84.

Alesina, Alberto, Rafael Di Tella, and Robert MacCulloch. "Inequality and Happiness: Are Europeans and Americans Different?" *Journal of Public Economics* 88, nos. 9–10 (2004): 2009–42.

Alesina, Alberto, Edward Glaeser, and Bruce Sacerdote. "Why Doesn't the United States Have a European-Style Welfare State?" *Brookings Papers on Economic Activity* 2 (2001).

Aletras, Nikolas, Dimitrios Tsarapatsanis, Daniel Preoţiuc-Pietro, and Vasileios Lampos. "Predicting Judicial Decisions of the European Court of Human Rights: A Natural Language Processing Perspective." *PeerJ Computer Science* 2, no. 93 (2016).

Allen, Robert. "The Industrial Revolution in Miniature: The Spinning Jenny in Britain, France, and India." Oxford University Working Paper No. 375 (2017).

Alstadsæter, Annette, Niels Johannesen, and Gabriel Zucman. "Tax Evasion and Inequality." *American Economic Review* 109, no. 6 (2019): 2073–103.

———. "Who Owns the Wealth in Tax Havens? Macro Evidence and Implications for Global Inequality." *Journal of Public Economics* 162 (2018): 89–100.

Alvaredo, Facundo, Lucas Chancel, Thomas Piketty, et al. *World Inequality Report*. Creative Commons, 2018.

Antràs, Pol, and Hans-Joachim Voth. "Factor Prices and Productivity Growth During the British Industrial Revolution." *Explorations in Economic History* 40 (2003): 52–77.

Arendt, Hannah. *The Human Condition*. London: University of Chicago Press, 1998.

Arthur, James, Kristján Kristjánsson, David Walker, et al. "Character Education in UK Schools Research Report." The Jubilee Centre for Character and Virtues at the University of Birmingham (2015).

Atkinson, Anthony B. *The Changing Distribution of Earnings in OECD Countries*. Oxford: Oxford University Press, 2009.

———. *Inequality: What Can Be Done?* London: Harvard University Press, 2015.

———. "The Restoration of Welfare Economics." *American Economic Review* 101, no. 3 (2011): 157–61.

Autor, David. "The Limits of the Digital Revolution: Why Our Washing Machines Won't Go to the Moon." Social Europe (2015), https://www.socialeurope.eu/the-limits-of-the-digital-revolution-why-our-washing-machines-wont-go-to-the-moon.

———. "Polanyi's Paradox and the Shape of Employment Growth." In "Re-evaluating Labor Market Dynamics: A Symposium Sponsored by the Federal Reserve Bank of Kansas City. Jackson Hole, Wyoming, August 21–23, 2014" (2015).

———. "The Polarization of Job Opportunities in the U.S. Labor Market: Implications for Employment and Earnings." Center for American Progress, April 2010.

———. "Skills, Education, and the Rise of Earnings Inequality Among the 'Other 99 Percent,'" *Science* 344, no. 6186 (2014): 843–51.

———. "Why Are There Still So Many Jobs? The History and Future of Workplace Automation." *Journal of Economics Perspectives* 29, no. 3 (2015): 3–30.

———. "Work of the Past, Work of the Future." Richard T. Ely Lecture delivered at the Annual Meeting of the American Economic Association (2019).

Autor, David, and David Dorn. "The Growth of Low-Skill Service Jobs and the Polarization of the US Labor Market." *American Economic Review* 103, no. 5 (2013): 1553–97.

———. "Technology Anxiety Past and Present." Bureau for Employers' Activities, International Labour Office (2013).

Autor, David, David Dorn, Lawrence Katz, et al. "The Fall of the Labor Share and the Rise of Superstar Firms." NBER Working Paper No. 23396 (2017).

Autor, David, Lawrence Katz, and Alan Krueger. "Computing Inequality: Have Computers Changed the Labour Market?" *Quarterly Journal of Economics* 133, no. 1 (1998): 1169–1213.

Autor, David, Frank Levy, and Richard Murnane. "The Skill Content of Recent Technological Change: An Empirical Exploration." *Quarterly Journal of Economics* 118, no. 4 (2003): 129–333.

Autor, David, and Anna Salomons. "Does Productivity Growth Threaten Employment? 'Robocalypse Now?'" Presentation at the European Central Bank Annual Conference (2017).

Avent, Ryan. *The Wealth of Humans: Work and Its Absence in the 21st Century*. London: Allen Lane, 2016.

Balme, Maurice. "Attitudes to Work and Leisure in Ancient Greece." *Greece & Rome* 31, no. 2 (1984): 140–52.

Barkai, Simcha. "Declining Labor and Capital Shares." Working Paper, University of Chicago (2016).

Beaudry, Paul, David Green, and Benjamin Sand. "The Great Reversal in the Demand for Skill and Cognitive Tasks." *Journal of Labor Economics* 34, no. 1 (2016): 199–247.

Becker, Gary. "The Economic Way of Looking at Life." Nobel Prize lecture, 9 December 1992.

Belfield, Chris, Claire Crawford, and Luke Sibieta. "Long-Run Comparisons for Spending per Pupil Across Different Stages of Education." Institute for Fiscal Studies, 27 February 2017.

Bell, Alex, Raj Chetty, Xavier Jaravel, et al. "Who Becomes an Inventor in America? The Importance of Exposure to Innovation." NBER Working Paper No. 24062 (2017).

Bell, Daniel. "The Bogey of Automation." *New York Review of Books*, 26 August 1965.

Berger, Thor, and Carl Frey. "Industrial Renewal in the 21st Century: Evidence from US Cities." *Regional Studies* (2015).

Berlin, Isaiah, *The Hedgehog and the Fox*. New York: Simon & Schuster, 1953.

———. *Two Concepts of Liberty*. Oxford: Clarendon Press, 1958.

Berman, Eli, John Bound, and Stephen Machin. "Implications of Skill-Biased Technological Change: International Evidence." *Quarterly Journal of Economics* 113, no. 4 (1998): 1245–79.

Bessen, James. "Toil and Technology." *IMF Financial and Development* 51, no. 1 (2015).

Beveridge, William. *Social Insurance and Allied Services*. London: Her Majesty's Stationery Office, 1942.

Birdwell, Jonathan, Ralph Scott, and Louis Reynolds. *Character Nation*. London: Demos, 2015.

Bloom, Benjamin. "The 2 Sigma Problem: The Search for Methods of Group Instruction as Effective as One-to-One Tutoring." *Educational Researcher* 13, no. 6 (1984): 4–16.

Boden, Margaret. *Philosophy of Artificial Intelligence*. Oxford: Oxford University Press, 1990.

Bostrom, Nick. "Ethical Issues in Advanced Artificial Intelligence." In George Lasker,

Wendell Wallach, and Iva Smit, eds., *Cognitive, Emotive, and Ethical Aspects of Decision Making in Humans and in Artificial Intelligence* (International Institute of Advanced Studies in Systems Research and Cybernetics, 2003), pp. 12–17.

Bostrom, Nick, and Eliezer Yudkowsky. "The Ethics of Artificial Intelligence." In William Ramsey and Keith Frankish, eds., *Cambridge Handbook of Artificial Intelligence* (Cambridge: Cambridge University Press, 2011).

Bricker, Jesse, Lisa J. Dettling, Alice Henriques, et al. "Changes in U.S. Family Finances from 2013 to 2016: Evidence from the Survey of Consumer Finances." *Federal Reserve Bulletin* 103:3 (2017).

Broadberry, Stephen, Bruce Campbell, Alexander Klein, et al. *British Economic Growth, 1270–1870*. Cambridge: Cambridge University Press, 2015.

Broudy, Eric. *The Book of Looms*. Hanover, NH: University Press of New England, 1979.

Brown, Noam, and Tuomas Sandholm. "Superhuman AI for Multiplayer Poker." *Science*, 11 July 2019.

Brynjolfsson, Erik. "AI and the Economy." Lecture at the Future of Life Institute, 1 July 2017.

Brynjolfsson, Erik, and Andrew McAfee. *The Second Machine Age*. London: W. W. Norton, 2014.

Brynjolfsson, Erik, and Tom Mitchell. "What Can Machine Learning Do? Workforce Implications." *Science* 358, no. 6370 (2017).

Caines, Colin, Florian Hoffman, and Gueorgui Kambourov. "Complex-Task Biased Technological Change and the Labor Market." International Finance Division Discussion Papers 1192 (2017).

Campbell, Murray A., Joseph Hoane Jr., and Feng-hsiung Hsu. "Deep Blue." *Artificial Intelligence* 134 (2002): 57–82.

Caplan, Bryan. *The Case Against Education: Why the Education System Is a Waste of Time and Money*. Oxford: Princeton University Press, 2018.

Carranza-Rojas, Jose, Herve Goeau, and Pierre Bonnet. "Going Deeper in the Automated Identification of Herbarium Specimens." *BMC Evolutionary Biology* 17, no. 181 (2017).

Chi Dao, Mai, Mitali Das, Zsoka Koczan, and Weicheng Lian. "Drivers of Declining Labor Share of Income." *IMF Blog* (2017)

Chien, YiLi, and Paul Morris. "Is U.S. Manufacturing Really Declining?" *Federal Bank of St. Louis Blog*, 11 April 2017.

Chui, Michael, Katy George, James Manyika, and Mehdi Miremadi. "Human + Machine: A New Era of Automation in Manufacturing." McKinsey & Co., September 2017.

Cingano, Federico, "Trends in Income Inequality and Its Impact on Economic Growth." OECD Social, Employment and Migration Working Paper No. 163 (2014).

Clark, Gregory, "The Agricultural Revolution and the Industrial Revolution: England, 1500–1912." Unpublished manuscript, University of California, Davis, 2002.

———. *A Farewell to Alms*. Princeton, NJ: Princeton University Press, 2007.

Cobb, Charles, and Paul Douglas. "A Theory of Production." *American Economic Review* 18, no. 1 (1928): 139–65.

Cohen, G. A. *If You're an Egalitarian, How Come You're So Rich?* London: Harvard University Press, 2001.

———. *Karl Marx's Theory of History: A Defence*. Oxford: Clarendon Press, 1978.

———. *Rescuing Justice and Equality*. London: Harvard University Press, 2008.

Cole, G. D. H. *A History of Socialist Thought*. London: St. Martin's Press, 1953.

Colton, Simon, and Geraint Wiggins. "Computational Creativity: The Final Frontier?" *Proceedings of the 20th European Conference on Artificial Intelligence* (2012), 21–26.

Cowen, Tyler. *Average Is Over: Powering America Beyond the Age of the Great Stagnation*. New York: Dutton, 2013.

Cox, Michael. "Schumpeter in His Own Words." *Federal Reserve Bank of Dallas: Economic Insights* 6, no. 3 (2001).

Crevier, Daniel. *AI: The Tumultuous History of the Search for Artificial Intelligence*. New York: Basic Books, 1993.

Cribb, Jonathan, Andrew Hood, Robert Joyce, and Agnes Norris Keller. "Living Standards, Poverty and Inequality in the UK: 2017." Institute for Fiscal Studies, 19 July 2017.

Dabla-Norris, Era, Kalpana Kochhar, Frantisek Ricka, et al. "Causes and Consequences of Income Inequality: A Global Perspective." IMF Staff Discussion Note (2015).

Darwin, Charles. *On the Origin of Species*. London: Penguin Books, 2009.

———. *Voyage of the Beagle*. London: Penguin Books, 1989.

Davis, Abe, Michael Rubinstein, Neal Wadhwa, et al. "The Visual Microphone: Passive Recovery of Sound from Video." *ACM Transactions on Graphics (TOG)* 33, no. 4 (2014).

Dawkins, Richard. *The Blind Watchmaker*. London: Penguin Books, 2016.

De Fauw, Jeffrey, Joseph Ledsam, Bernardino Romera-Paredes, et al. "Clinically Applicable Deep Learning for Diagnosis and Referral in Retinal Disease." *Nature Medicine* 24 (2018): 1342–50.

Deloitte. "From Brawn to Brains: The Impact of Technology on Jobs in the UK" (2015).

Deming, David. "The Growing Importance of Social Skills in the Labor Market." *Quarterly Journal of Economics* 132, no. 4 (2017): 1593–1640.

Dennett, Daniel. *From Bacteria to Bach and Back*. London: Allen Lane, 2017.

———. "A Perfect and Beautiful Machine: What Darwin's Theory of Evolution Reveals About Artificial Intelligence." *Atlantic*, 22 June 2012.

Diamond, Peter, and Emmanuel Saez. "The Case for a Progressive Tax: From Basic Research to Policy Recommendations." *Journal of Economic Perspectives* 25, no. 4 (2011): 165–90.

Dimsdale, Nicholas, Nicholas Horsewood, and Arthur Van Riel. "Unemployment in Interwar Germany: An Analysis of the Labor Market, 1927–1936." *Journal of Economic History* 66, no. 3 (2006): 778–808.

Dreyfus, Hubert. *What Computers Can't Do: The Limits of Artificial Intelligence*. New York: Harper & Row, 1979.

Eberstadt, Nicholas. *Men Without Work: America's Invisible Crisis*. West Conshohocken, PA: Templeton Press, 2016.

Eliot, T. S. *Collected Poems 1909–1962*. London: Faber and Faber, 2002.

Elliott, Stuart W. "Computers and the Future of Skill Demand." *OECD Educational Research and Innovation* (2017).

Elster, Jon. "Comment on Van der Veen and Van Parijs." *Theory and Society* 15, no. 5 (1986): 709–21.

Esteva, Andre, Brett Kuprel, Roberto A. Novoa, et al. "Dermatologist-Level Classification of Skin Cancer with Deep Neural Networks." *Nature* 542 (2017): 115–18.

Executive Office of the President. *Artificial Intelligence, Automation, and the Economy* (December 2016).

Ezrachi, Ariel, and Maurice Stucke. *Virtual Competition: The Promise and Perils of the Algorithm-Driven Economy*. Cambridge, MA: Harvard University Press, 2016.

Feigenbaum, Gustavo. *Conversations with John Searle*. Buenos Aires: Libros En Red, 2003.

Felipe, Jesus, Connie Bayudan-Dacuycuy, and Matteo Lanzafame. "The Declining Share of Agricultural Employment in China: How Fast?" *Structural Change and Economic Dynamics* 37 (2016): 127–37.

Frayne, David. *The Refusal of Work: The Theory and Practice of Resistance to Work*. London: Zed Books, 2015.

Freud, Sigmund. *Civilization and Its Discontents*. New York: W. W. Norton, 2010.

Freund, Caroline, and Sarah Oliver. "The Origins of the Superrich: The Billionaire Characteristics Database." *Peterson Institute for International Economics* 16, no. 1 (2016).

Frey, Carl, and Michael Osborne. "The Future of Employment: How Susceptible Are Jobs to Computerisation?" *Technological Forecasting and Social Change* 114 (January 2017): 254–80.

Frey, Carl, Michael Osborne, Craig Holmes, et al. "Technology at Work v2.0: The Future Is Not What It Used to Be." Oxford Martin School and Citi (2016).

Friedman, Benjamin M. "Born to Be Free." *New York Review of Books*, 12 October 2017.

Fromm, Eric. *Fear of Freedom*. Abingdon: Routledge, 2009.

Fry, Hannah. *Hello World: How to Be Human in the Age of the Machine*. London: Penguin, 2018.

Galbraith, John Kenneth. *The Affluent Society*. London: Penguin Books, 1999.

———. *American Capitalism: The Concept of Countervailing Power*. Eastford, CT: Martino Fine Books, 2012.

Garber, Marjorie. *Vested Interests: Cross-Dressing and Cultural Anxiety*. New York: Routledge, 2012.

Gerth, Hans H., and C. Wright Mills, eds. "Introduction: The Man and His Work." In *From Max Weber: Essays in Sociology*. Oxford: Oxford University Press, 1946.

Goldin, Claudia, and Lawrence Katz. *The Race Between Education and Technology*. London: Harvard University Press, 2009.

Good, Irving John. "Speculations Concerning the First Ultraintelligent Machine." *Advances in Computers* 6 (1966): 31–88.

Goodman, Joshua, Julia Melkers, and Amanda Pallais. "Can Online Delivery Increase Access to Education?" *Journal of Labor Economics* 37, no. 1 (2019).

Goos, Maarten, and Alan Manning. "Lousy and Lovely Jobs: The Rising Polarization of Work in Britain." *Review of Economics and Statistics* 89, no. 1 (2007): 119–33.

Goos, Maarten, Alan Manning, and Anna Salomons. "Explaining Job Polarization: Routine-Biased Technological Change and Offshoring." *American Economic Review* 104, no. 8 (2014): 2509–26.

Gordon, Robert. *The Rise and Fall of American Growth*. Oxford: Princeton University Press, 2017.

Grace, Katja, John Salvatier, Allan Dafoe, et al. "When Will AI Exceed Human Performance? Evidence from AI Experts." *Journal of Artificial Intelligence Research* 62 (2018): 729–54.

Graetz, Georg, and Guy Michaels. "Robots at Work." *Review of Economics and Statistics* 100, no. 5 (2018): 753–68.

Grossman, Maura, and Gordon Cormack. "Technology-Assisted Review in e-Discovery Can Be More Effective and More Efficient than Exhaustive Manual Review." *Richmond Journal of Law and Technology* 17, no. 3 (2011).

Haldane, Andy. "In Giving, How Much Do We Receive? The Social Value of Volunteering." Lecture to the Society of Business Economists, London, 9 September 2014.

———. "Labour's Share." Speech at the Trades Union Congress, London, 12 November 2015.

Harari, Yuval Noah. *Homo Deus: A Brief History of Tomorrow*. London: Harvill Secker, 2016.

———. *Sapiens*. London: Harvill Secker, 2011.

Harris, Sam. "Can We Build AI Without Losing Control over It?" TED talk, 29 September 2016.

Harrison, Mark. "Soviet Economic Growth Since 1928: The Alternative Statistics of G. I. Khanin." *Europe–Asia Studies* 45, no. 1 (1993): 141–67.

Hassabis, Demis. "Artificial Intelligence: Chess Match of the Century." *Nature* 544 (2017): 413–14.

Haugeland, John. *Artificial Intelligence: The Very Idea*. London: MIT Press, 1989.

Hawking, Stephen. *On the Shoulders of Giants: The Great Works of Physics and Astronomy*. London: Penguin, 2003.

Hesiod. *Theogony, Works and Days, Testimonia*, Glenn W. Most, ed. and trans. Loeb Classical Library 57. London: Harvard University Press, 2006.

Hobsbawm, Eric. *Industry and Empire*. London: Penguin, 1999.

Hofstadter, Douglas. *Gödel, Escher, Bach: An Eternal Golden Braid*. London: Penguin, 2000.

———. "Just Who Will Be We, in 2493?" Indiana University, Bloomington (2003).

———. "Staring Emmy Straight in the Eye—And Doing My Best Not to Flinch." In David Cope, ed., *Virtual Music: Computer Synthesis of Musical Style*. London: MIT Press, 2004.

Holtz-Eakin, Douglas, David Joulfaian, and Harvey Rosen. "The Carnegie Conjecture: Some Empirical Evidence." *Quarterly Journal of Economics* 108, no. 2 (1993): 413–35.

Hughes, Chris. *Fair Shot: Rethinking Inequality and How We Earn*. London: Bloomsbury, 2018.

Imbens, Guido, Donald Rubin, and Bruce Sacerdote. "Estimating the Effect of Unearned Income on Labor Earnings, Savings, and Consumption: Evidence from a Survey of Lottery Players." *American Economic Review* 91, no. 4 (2001): 778–94.

IMF. *World Economic Outlook* (2017)

International Association of Machinists. "Workers' Technology Bill of Rights." *Democracy* 3, no. 1 (1983): 25–27.

International Labour Organization. *Care Work and Care Jobs for the Future of Decent Work*. Geneva: International Labour Office, 2018.

———. *Global Wage Report 2014/2015*. Geneva: International Labour Office, 2015.

Jahoda, Marie. *Employment and Unemployment: A Social-Psychological Analysis*. Cambridge: Cambridge University Press, 1982.

Jahoda, Marie, Paul Lazarsfeld, and Hans Zeisel. *Marienthal: The Sociography of an Unemployed Community*. Piscataway, NJ: Transaction Publishers, 2009.

Johnston, David. *Equality*. Indianapolis: Hackett Publishing, 2000.

Jones, Charles I. "The Facts of Economic Growth." In John B. Taylor and Harald Uhlig, eds., *Handbook of Macroeconomics*, vol. 2A, pp. 3–69. Amsterdam: Elsevier, 2016.

Jones, Damon, and Ioana Marinescu. "The Labor Market Impact of Universal and Permanent Cash Transfers: Evidence from the Alaska Permanent Fund." NBER Working Paper No. 24312 (February 2018).

Kaldor, Nicholas. "A Model of Economic Growth." *Economic Journal* 67:268 (1957), 591–624.

Kalokerinos, Elise, Kathleen Kjelsaas, Steven Bennetts, and Courtney von Hippel. "Men in Pink Collars: Stereotype Threat and Disengagement Among Teachers and Child Protection Workers." *European Journal of Social Psychology* 47, no. 5 (2017).

Karabarbounis, Loukas, and Brent Neiman. "The Global Decline of the Labor Share." *Quarterly Journal of Economics* 129, no. 1 (2014): 61–103.

Kasparov, Garry. "The Chess Master and the Computer." *New York Review of Books*, 11 February 2010.

———. *Deep Thinking*. London: John Murray, 2017.

Katz, Daniel Marin, Michael J. Bommarito II, and Josh Blackman. "A General Approach for Predicting the Behavior of the Supreme Court of the United States." *PLOS ONE*, 12 April 2017.

Keynes, John Maynard. *Essays in Persuasion*. New York: W. W. Norton, 1963.

———. "Relative Movements of Real Wages and Output." *Economic Journal* 49, no. 93 (1939): 34–51.

Khan, Lina M. "Amazon's Antitrust Paradox." *Yale Law Journal* 126, no. 3 (2017): 564–907.

Kheraj, Sean. "The Great Epizootic of 1872–73: Networks of Animal Disease in North American Urban Environments." *Environmental History* 23, no. 3 (2018).

Kolbert, Elizabeth. "Hosed: Is There a Quick Fix for the Climate?" *New Yorker*, 8 November 2009.

Krämer, Hagen. "Bowley's Law: The Diffusion of an Empirical Supposition into Economic Theory." *Papers in Political Economy* 61 (2011).

Kymlicka, Will. *Contemporary Political Philosophy: An Introduction*. New York: Oxford University Press, 2002.

Landes, David. *Abba Ptachya Lerner 1903–1982: A Biographical Memoir*. Washington, DC: National Academy of Sciences, 1994.

Lay, Maxwell. *Ways of the World: A History of the World's Roads and of the Vehicles That Used Them*. New Brunswick, NJ: Rutgers University Press, 1992.

Le, Quoc, Marc'Aurelio Ranzato, Rajat Monga, et al. "Building High-Level Features Using Large Scale Unsupervised Learning." *Proceedings of the 29th International Conference on Machine Learning* (2012).

Leontief, Wassily. "Is Technological Unemployment Inevitable?" *Challenge* 22, no. 4 (1979): 48–50.

———. "National Perspective: The Definition of Problems and Opportunities." In *The Long-Term Impact of Technology on Employment and Unemployment: A National Academy of Engineering Symposium, 30 June 1983*. Washington, DC: National Academy Press, 1983.

———. "Technological Advance, Economic Growth, and the Distribution of Income." *Population and Development Review* 9, no. 3 (1983): 403–10.

Levitt, Steven, and Stephen Dubner. *Superfreakonomics*. New York: HarperCollins, 2009.

Lindley, Joanne, and Stephen Machin. "The Rising Postgraduate Wage Premium." *Economica* 83 (2016): 281–306.

Longmate, Norman. *The Workhouse: A Social History*. London: Pimlico, 2003.

Lovelock, James. *Novacene*, London: Allen Lane, 2019.

Lowrey, Annie. *Give People Money: The Simple Idea to Solve Inequality and Revolutionise Our Lives*. London: W. H. Allen, 2018.

Luce, Edward. *The Retreat of Western Liberalism*. London: Little, Brown, 2017.

Mankiw, Nicholas G. "Yes, $r > g$. So What?" *American Economic Review: Papers & Proceedings* 105, no. 5 (2015): 43–47.

Manuelli, Rodolfo, and Ananth Seshadri. "Frictionless Technology Diffusion: The Case of Tractors." *American Economic Review* 104, no. 4 (2014): 1268–1391.

Marovčík, Matej, Martin Schmid, Neil Burch, et al. "Deep Stack: Expert-Level Artificial Intelligence in Heads-Up No-Limit Poker." *Science* 356, no. 6337 (2017): 508–13.

Marr, David. *Vision: A Computational Investigation into the Human Representation and Processing of Visual Information*. London: MIT Press, 2010.

Marshall, Alfred. *Principles of Economics*. London: Macmillan, 1890.

Marx, Karl. *Selected Writings*, Simon Lawrence, ed. Indianapolis: Hackett, 1994.

McCarthy, John, Marvin Minsky, Nathaniel Rochester, and Claude Shannon. "A Proposal for the Dartmouth Summer Research Project on Artificial Intelligence." 31 August 1955.

McCulloch, Warren, and Walter Pitts. "A Logical Calculus of the Ideas Immanent in Nervous Activity." *Bulletin of Mathematical Biophysics* 5 (1943): 115–33.

McKinsey Global Institute. "A Future That Works: Automation, Employment, and Productivity." January 2017.

Mill, John Stuart. *Principles of Political Economy with Chapters on Socialism*. Oxford: Oxford University Press, 2008.

———. *Principles of Political Economy with Some of Their Applications to Social Philosophy*. London: Longmans, Green, 1848.

Minsky, Marvin. "Neural Nets and the Brain Model Problem." PhD diss., Princeton University, 1954.

———. *Semantic Information Processing*. Cambridge, MA: MIT Press, 1968.

Mirrlees, James, and Stuart Adam. *Dimensions of Tax Design: The Mirrlees Review*. Oxford: Oxford University Press, 2010.

Mnih, Volodymyr, Koray Kavukcuoglu, David Silver, et al. "Human-Level Control Through Deep Reinforcement Learning." *Nature* 518 (25 February 2015): 529–33.

Mokyr, Joel. *The Lever of Riches: Technological Creativity and Economic Progress*. New York: Oxford University Press, 1990.

———. "Technological Inertia in Economic History." *Journal of Economic History* 52, no. 2 (1992): 325–38.

Mokyr, Joel, Chris Vickers, and Nicholas Ziebarth. "The History of Technological Anxiety and the Future of Economic Growth: Is This Time Different?" *Journal of Economic Perspectives* 29, no. 3 (2015): 31–50.

Moravec, Hans. *Mind Children*. Cambridge, MA: Harvard University Press, 1988.

Moretti, Enrico. *The New Geography of Jobs*. New York: First Mariner Books, 2013.

Morozov, Evgeny. *To Save Everything, Click Here: Technology, Solutionism, and the Urge to Fix Problems That Don't Exist.* New York: PublicAffairs, 2013.
Motta, Massimo. *Competition Policy.* Cambridge: Cambridge University Press, 2007.
Müller, Karsten, and Carlo Schwarz. "Fanning the Flames of Hate: Social Media and Hate Crime." Warwick University Working Paper Series No. 373 (May 2018).
Newell, Alan, and Herbert Simon. "GPS, A Program That Simulates Human Thought." In H. Billing, ed., *Lernende automaten.* Munich: R. Oldenbourgh, 1961.
Ng, Andrew. "What Artificial Intelligence Can and Can't Do Right Now." *Harvard Business Review,* 9 November 2016.
Nilsson, Nils J. "Artificial Intelligence, Employment, and Income." *AI Magazine* (Summer 1984).
———. *The Quest for Artificial Intelligence* (New York: Cambridge University Press, 2010).
Nordhaus, William. "Two Centuries of Productivity Growth in Computing." *Journal of Economic History* 67, no. 1 (2007): 128–59.
Novak, David. "Toward a Jewish Public Philosophy in America." In Alan Mittleman, Robert Licht, and Jonathan D. Sarna, eds., *Jews and the American Public Square: Debating Religion and Republic.* Lanham, MD: Rowman & Littlefield, 2002.
Nübler, Irmgard. "New Technologies: A Jobless Future or Golden Age of Job Creation?" International Labour Office Working Paper No. 13 (2016).
OECD. "Divided We Stand: Why Inequality Keeps Rising." 2011.
———. "Focus on Top Incomes and Taxation in OECD Countries: Was the Crisis a Game Changer?" May 2014.
———. "Growing Income Inequality in OECD Countries: What Drives It and How Can Policy Tackle It?" 2011.
———. *Health at a Glance 2017: OECD Indicators.* Paris: OECD Publishing, 2018.
———. "Hours Worked: Average Annual Hours Actually Worked." OECD Employment and Labour Market Statistics database, 2019.
———. *Job Creation and Local Economic Development 2018: Preparing for the Future of Work.* Paris: OECD Publishing, 2018.
———. *OECD Employment Outlook 2012.* Paris: OECD Publishing, 2012.
———. *OECD Employment Outlook 2016,* no. 1. Paris: OECD Publishing, 2016.
———. *OECD Employment Outlook 2017.* Paris: OECD Publishing, 2017.
———. *OECD Employment Outlook 2018.* Paris: OECD Publishing, 2018.
Oliveira, Victor. "The Food Assistance Landscape." Economic Research Service at the United States Department of Agriculture, Economic Information Bulletin Number 169 (March 2017).
O'Neil, Cathy. *Weapons of Math Destruction: How Big Data Increases Inequality and Threatens Democracy.* New York: Crown, 2016.
O'Rourke, Kevin, Ahmed Rahman, and Alan Taylor. "Luddites, the Industrial Revolution, and the Demographic Transition." *Journal of Economic Growth* 18, no. 4 (2013): 373–409.
Orwell, George. *Essays.* London: Penguin Books, 2000.
Ostry, Jonathan, Andrew Berg, and Charalambos Tsangarides. "Redistribution, Inequality, and Growth." IMF Staff Discussion Note (February 2014).
Paine, Thomas, *Agrarian Justice.* Digital edition, 1999.

Paley, William. *Natural Theology*. Oxford: Oxford University Press, 2008.
Pigou, Arthur. *A Study in Public Finance*. London: Macmillan, 1928.
Piketty, Thomas. *Capital in the Twenty-First Century*. London: Harvard University Press, 2014.
Piketty, Thomas, and Emmanuel Saez. "A Theory of Optimal Capital Taxation." NBER Working Paper No. 17989 (2012).
Piketty, Thomas, Emmanuel Saez, and Gabriel Zucman. "Distribution National Accounts: Methods and Estimates for the United States." *Quarterly Journal of Economics* 133, no. 2 (2018): 553–609.
Piketty, Thomas, and Gabriel Zucman. "Capital Is Back: Wealth–Income Ratios in Rich Countries 1700–2010." *Quarterly Journal of Economics* 129, no. 3 (2014): 1255–1310.
Pleijt, Alexandra, and Jacob Weisdorf. "Human Capital Formation from Occupations: The 'Deskilling Hypothesis' Revisited." *Cliometrica* 11, no. 1 (2017): 1–30.
Polanyi, Michael. *The Tacit Dimension*. Chicago: Chicago University Press, 1966.
Popper, Karl. *The Open Society and Its Enemies*, vol. 1: *The Age of Plato*. London: Routledge, 1945.
Putnam, Hilary. "Much Ado About Not Very Much." *Daedalus* 117, no. 1 (1988): 269–81.
PwC. "Global Top 100 Companies by Market Capitalisation." 2018.
———. "Workforce of the Future: The Competing Forces Shaping 2030." 2018.
Rawls, John. *A Theory of Justice*. Cambridge, MA: Harvard University Press, 1999.
Reichinstein, David. *Albert Einstein: A Picture of His Life and His Conception of the World*. Prague: Stella Publishing House, 1934.
Remus, Dana, and Frank Levy. "Can Robots Be Lawyers? Computers, Lawyers, and the Practice of Law." *Georgetown Journal of Legal Ethics* 30, no. 3 (2017): 501–58.
Renshaw, James. *In Search of the Greeks*. 2nd edition. London: Bloomsbury, 2015.
Ricardo, David. *Principles of Political Economy and Taxation*. New York: Prometheus Books, 1996.
Ruger, Theodore W., Pauline T. Kim, Andrew D. Martin, and Kevin M. Quinn. "The Supreme Court Forecasting Project: Legal and Political Science Approaches to Predicting Supreme Court Decisionmaking." *Columbia Law Review* 104, no. 4 (2004): 1150–1210.
Russakovsky, Olga, Jia Deng, Hao Su, et al. "ImageNet Large Scale Visual Recognition Challenge." *International Journal of Computer Vision* 115, no. 3 (2015): 211–52.
Russell, Bertrand. *In Praise of Idleness and Other Essays*. New York: Routledge, 2004.
Saez, Emmanuel. "Striking It Richer: The Evolution of Top Incomes in the United States." Published online at https://eml.berkeley.edu/~saez/ (2016).
Saez, Emmanuel, and Thomas Piketty. "Income Inequality in the United States, 1913–1998." *Quarterly Journal of Economics* 118, no. 1 (2003), 1–39.
Saez, Emmanuel, and Gabriel Zucman. "Wealth Inequality in the United States Since 1913: Evidence from Capitalized Income Tax Data." *Quarterly Journal of Economics* 131, no. 2 (2016): 519–78.
Salehi-Isfahani, Djaved, and Mohammad Mostafavi-Dehzooei. "Cash Transfers and Labor Supply: Evidence from a Large-Scale Program in Iran." *Journal of Development Economics* 135 (2018): 349–67.
Sandel, Michael. "In Conversation with Michael Sandel: Capitalism, Democracy, and the Public Good." LSE Public Lecture chaired by Tim Besley, 2 March 2017.

———. "Themes of 2016: Progressive Parties Have to Address the People's Anger." *Guardian*, 1 January 2017.

Scanlon, Tim, *Why Does Inequality Matter?* Oxford: Oxford University Press, 2018.

Schaff, Kory. *Philosophy and the Problems of Work: A Reader*. Oxford: Rowman & Littlefield, 2001.

Scheidel, Walter. *The Great Leveler: Violence and the History of Inequality from the Stone Age to the Twenty-First Century*. Oxford: Princeton University Press, 2017.

Schloss, David. *Methods of Industrial Remuneration*. London: Williams and Norgate, 1898.

Schumpeter, Joseph A. *Capitalism, Socialism, and Democracy*. London: Routledge, 2005.

Searle, John. "Minds, Brains, and Programs." *Behavioral and Brain Sciences* 3 (1980): 417–57.

———. "Watson Doesn't Know It Won on 'Jeopardy!'" *Wall Street Journal*, 23 February 2011.

Selbst, Andrew, and Julia Powles. "Meaningful Information and the Right to Explanation." *International Data Privacy Law* 7, no. 4 (2017): 233–42.

Seligman, Ben. *Most Notorious Victory: Man in an Age of Automation*. New York: Free Press, 1966.

Shapiro, H. A. "'Heros Theos': The Death and Apotheosis of Herakles." *Classical World* 77, no. 1 (1983): 7–18.

Silk, Leonard. "Economic Scene; Structural Joblessness." *New York Times*, 6 April 1983.

Silver, David, Aja Huang, Chris Maddison, et al. "Mastering the Game of Go with Deep Neural Networks and Tree Search." *Nature* 529 (2016): 484–89.

Silver, David, Thomas Hubert, Julian Schrittwieser, et al. "Mastering Chess and Shogi by Self-Play with a General Reinforcement Learning Algorithm." https://arxiv.org, arXiv:1712.01815v1 (2017)

Silver, David, Julian Schrittwieser, Karen Simonyan, et al. "Mastering the Game of Go Without Human Knowledge." *Nature* 550 (2017), 354–9.

Singh, Satinder, Andy Okun, and Andrew Jackson. "Artificial Intelligence: Learning to Play Go from Scratch." *Nature* 550 (2017): 336–37.

Smith, Adam. *An Inquiry into the Nature and Causes of the Wealth of Nations*. Oxford: Oxford University Press, 1998.

———. *The Theory of Moral Sentiments*. London: Penguin Books, 2009.

Snyder, Jacob. "Leisure in Aristotle's Political Thought." *Polis: The Journal for Ancient Greek Political Thought* 35, no. (2018).

Solomonoff, Grace. "Ray Solomonoff and the Dartmouth Summer Research Project in Artificial Intelligence." No date. http://raysolomonoff.com/dartmouth/dartray.pdf.

Somers, James. "The Man Who Would Teach Machines to Think." *Atlantic*, November 2013.

Spencer, David. *The Political Economy of Work*. Digital edition. New York: Routledge, 2010.

Standage, Tom. *The Turk*. New York: Berkley Publishing Group, 2002.

Stiglitz, Joseph. "Inequality and Economic Growth." *Political Quarterly* 86, no. 1 (2016): 134–55.

———. "Towards a General Theory of Consumerism: Reflections on Keynes's Economic Possibilities for Our Grandchildren." In Lorenzo Pecchi and Gustavo Piga, eds.,

Revisiting Keynes: Economics Possibilities for Our Grandchildren. Cambridge, MA: MIT Press, 2008.

Summers, Lawrence. "The 2013 Martin Feldstein Lecture: Economic Possibilities for Our Children." *NBER Reporter* 4 (2013).

Susskind, Daniel. "Automation and Demand." Oxford University Department of Economics Discussion Paper Series No. 845 (2018).

———. "A Model of Technological Unemployment." Oxford University Department of Economics Discussion Paper Series No. 819 (2017).

———. "Re-Thinking the Capabilities of Technology in Economics." *Economics Bulletin* 39, no. 1 (2019): A30.

———. "Robots Probably Won't Take Our Jobs—For Now." *Prospect*, 17 March 2017.

———. "Technology and Employment: Tasks, Capabilities and Tastes." DPhil. diss. Oxford University, 2016.

———. "Three Myths About the Future of Work (and Why They Are Wrong)." TED talk, March 2018.

Susskind, Daniel, and Richard Susskind. *The Future of the Professions*. Oxford: Oxford University Press, 2015.

———. "The Future of the Professions." *Proceedings of the American Philosophical Society* (2018).

Susskind, Jamie. *Future Politics*. Oxford: Oxford University Press, 2018.

———. "Future Politics: Living Together in a World Transformed by Tech." Google Talks, 18 October 2018.

———. "Future Politics: Living Together in a World Transformed by Tech." Harvard University CLP Speaker Series, 11 December 2018.

Suzman, James, *Affluence Without Abundance: The Disappearing World of the Bushmen*. London: Bloomsbury, 2017.

Syverson, Chad. "Challenges to Mismeasurement Explanations for the US Productivity Slowdown." *Journal of Economic Perspectives* 32, no. 2 (2017): 165–86.

Taplin, Jonathan. "Is It Time to Break Up Google?" *New York Times*, 22 April 2017.

Tegmark, Max. *Life 3.0: Being Human in the Age of Artificial Intelligence*. London: Penguin Books, 2017.

Tennyson, Alfred. *Selected Poems*. London: Penguin Books, 2007.

Thiel, Peter, and Blake Masters. *Zero to One*. New York: Crown Business, 2014.

Tombs, Robert. *The English and Their History*. London: Penguin Books, 2015.

Topol, Eric, "High-Performance Medicine: The Convergence of Human and Artificial Intelligence." *Nature* 25 (2019): 44–56.

Turing, Alan. "Intelligent Machinery, A Heretical Theory." *Philosophia Mathematica* 3, no. 4 (1996): 156–260.

———. "Intelligence Machinery: A Report by A. M. Turing." National Physical Laboratory (1948). Archived at https://www.npl.co.uk/getattachment/about-us/History/Famous-faces/Alan-Turing/80916595-Intelligent-Machinery.pdf (accessed July 2018).

———. "Lecture to the London Mathematical Society." 20 February 1947.

Tyson, Laura, and Michael Spence. "Exploring the Effects of Technology on Income and Wealth Inequality." In Heather Boushey, J. Bradford DeLong, and Marshall Steinbaum, eds., *After Piketty: The Agenda for Economics and Inequality*. London: Harvard University Press, 2017.

UBS. "Intelligence Automation: A UBS Group Innovation White Paper." 2017.

Van Parijs, Phillippe. "Basic Income: A Simple and Powerful Idea for the Twenty-First Century." In Bruce Ackerman, Anne Alstott, and Phillipe Van Parijs, eds., *Redesigning Distribution: Basic Income and Stakeholder Grants as Cornerstones for an Egalitarian Capitalism*. New York: Verso, 2005.

Van Parijs, Phillippe, and Yannick Vanderborght. *Basic Income: A Radical Proposal for a Free Society and a Sane Economy*. London: Harvard University Press, 2017.

van Zanden, Jan Luiten, Joerg Baten, Marco Mira d'Ercole, et al. "How Was Life? Global Well-Being Since 1820." Paris: OECD Publishing, 2014.

Veblen, Thorstein. *The Theory of the Leisure Class*. New York: Dover Thrift Editions, 1994.

Walker, Tom. "Why Economists Dislike a Lump of Labor." *Review of Social Economy* 65, no. 3 (2007): 279–91.

Wang, Dayong, Aditya Khosla, Rishab Gargeya, et al. "Deep Learning for Identifying Metastatic Breast Cancer." https://arxiv.org, arXiv:1606.05718 (2016).

Webb, Michael, Nick Short, Nicholas Bloom, and Josh Lerner. "Some Facts of High-Tech Patenting." NBER Working Paper No. 24793 (2018).

Weber, Bruce. "Mean Chess-Playing Computer Tears at Meaning of Thought." *New York Times*, 19 February 1996.

Weber, Max. *The Protestant Ethic and the Spirit of Capitalism*. Oxford: Oxford University Press, 2011.

Weil, David. *Economic Growth*. 3rd edition. London: Routledge, 2016.

Weiss, Antonia. "Harold Bloom, the Art of Criticism No. 1." *Paris Review* 118 (Spring 1991).

Weizenbaum, Joseph. *Computer Power and Human Reason*. San Francisco: W. H. Freeman, 1976.

———. "ELIZA—A Computer Program for the Study of Natural Language Communication Between Man and Machine." *Communications of the ACM* 9, no. 1 (1966): 36–45.

Wood, Gaby. *Living Dolls*. London: Faber and Faber, 2002.

World Bank. *World Development Report: Digital Dividends*. 2016.

World Economic Forum. *Global Risks Report 2017*. 2017.

Wu, Zhe, Bharat Singh, Larry S. Davis, and V. S. Subrahmanian. "Deception Detection in Videos." https://arxiv.org/abs/1712.04415, 12 September 2017.

Zucman, Gabriel. "Taxing Across Borders: Tracking Personal Wealth and Corporate Profits." *Journal of Economic Perspectives* 28, no. 4 (2014): 121–48.

Zweig, Stefan. *The World of Yesterday*. London: Pushkin Press, 2014.

ACKNOWLEDGMENTS

I wrote much of this book at Balliol College, Oxford. I would like to thank all my friends and colleagues there—in particular David Vines, James Forder, and Nicky Trott—who have made it such a happy and productive intellectual home from which to work. Balliol is a special place, and it has been a privilege to be part of the community for so long.

I would also like to thank my brilliant literary agents, Georgina Capel, Rachel Conway, and Irene Baldoni, for their support and encouragement. I have felt spoiled to work with two formidable editors in writing this book: Laura Stickney and Grigory Tovbis, who have been relentlessly helpful, insightful, and supportive. Thank you to the rest of the team as well—Holly Hunter, Isabel Blake, Will O'Mullane, Olivia Andersen, and Anna Hervé at Allen Lane, and Maggie Richards, Jessica Wiener, Carolyn O'Keefe, and Christopher O'Connell at Metropolitan. And particular thanks to Sara Bershtel, for her confidence in my work and for taking me on in the first place.

Special thanks must go to Daniel Chandler, Arthur Hughes-Hallett, and Tom Woodward, for reading and commenting on the manuscript so carefully and thoughtfully at various stages. Thank you to Alex Canfor-Dumas, Josh Glancy, and Owain Williams for so often discussing the ideas in this book with me over the years; to Jane Birdsell and Muriel Jorgensen, who copyedited the book and saved me from numerous embarrassments; and to Rebecca Clark for fact-checking the manuscript. And many thanks to all my other students—our conversations are an ongoing source of inspiration.

A very warm thank you to my parents-in-law, Thomas and Jules Hughes-Hallett, whose study in Suffolk became an empty-tea-cup-ridden writing refuge for me on countless stays.

The biggest thanks must go to my fellow Susskinds. My mum, Michelle, is my biggest supporter in life; her continual guidance and encouragement helped me so much in writing this book. My sister, Ali, was always on hand to offer her wisdom, advice, and inimitable humor. And I thank my brother, Jamie, the smartest and most capable person I know, who pored over these pages for me several times, and whose judgment I trust more than anyone's.

Then there is my dad, Richard. In his recent book, Jamie wrote that he did not have the words to describe what he owes to our dad and his gratitude to him. I feel much the same way. I was so hugely proud to write my first book with him. And though his name is not on the cover on this book, his influence is in every chapter, as will be plain to any reader. No son could have a better dad than mine, and I feel so lucky to have him in my corner. Thank you, dad.

And finally, Grace and Rosa, the two loves of my life. Thank you so much for being there with me along the way (Rosa, for the second half of the journey). This book, and everything that I do, is dedicated to you both.

INDEX

Pages numbers in *italics* refer to figures.